THE FRONTIERS
OF KNOWLEDGE

CONTRIBUTORS:

ISAAC ASIMOV   SIR PETER MEDAWAR

DANIEL BELL   AKIO MORITA

SAUL BELLOW   EDMUNDO O'GORMAN

ARTHUR C. CLARKE   MOSHE SAFDIE

SIR EDMUND HILLARY   JAMES D. WATSON

SIR FRED HOYLE   CASPAR W. WEINBERGER

WILLARD F. LIBBY   HUW WHELDON

THE FRONTIERS OF KNOWLEDGE

FOREWORD BY
BROOKE HINDLE

THE FRANK NELSON DOUBLEDAY
LECTURES
AT THE NATIONAL MUSEUM
OF HISTORY AND TECHNOLOGY
SMITHSONIAN INSTITUTION
WASHINGTON, D.C.

Doubleday & Company, Inc.

GARDEN CITY, NEW YORK

1975

The Frank Nelson Doubleday Lectures at the National Museum of History and Technology, Smithsonian Institution, Washington, D.C., are made possible by a continuing grant from Doubleday & Company, Inc., in honor of the founder of the publishing house.

An excerpt from Saul Bellow's lecture "Literature in the Age of Technology" appeared in the Washington *Post*, November 1972.

Daniel Bell's lecture "Technology, Nature, and Society" appeared in *The American Scholar*, June 1973 and was excerpted in the Washington *Post*, December 1973.

Sir Peter Medawar's lecture "Technology and Evolution" appeared in the *Smithsonian* magazine, May 1973.

James D. Watson's lecture "The Dissemination of Unpublished Information" was excerpted in the Washington *Post*, May 1974.

The first series of lectures, TECHNOLOGY AND THE FRONTIERS OF KNOWLEDGE, was published in a separate edition by Doubleday & Company, Inc., in 1975.

Library of Congress Cataloging in Publication Data

Main entry under title: The Frontiers of knowledge.

(The Frank Nelson Doubleday lectures)

CONTENTS: First series, 1972–73, Technology and the frontiers of knowledge: Bellow, S. Literature in the age of technology. Bell, D. Technology, nature, and society. O'Gorman, E. History, technology, and the pursuit of happiness. Medawar, P., Sir. Technology and evolution. Clarke, A. C. Technology and the limits of knowledge.—Second series, 1973–74, Creativity and collaboration: Morita, A. Creativity in modern industry. Watson, J. D. The dissemination of unpublished information. [etc.]

Includes bibliographical references.

1. Technology and civilization—Addresses, essays, lectures. I. Series.

CB478.F76 1975   909

ISBN 0-385-03151-3

Library of Congress Catalog Card Number 74–18793

FOREWORD

The Frank Nelson Doubleday Lectures, *The Frontiers of Knowledge*, express succinctly the mission of the Smithsonian Institution—"the increase and diffusion of knowledge." Each of the lecturers has contributed measurably to the advancement of at least one of our frontiers of knowledge. Some walked over ground where none had walked before; some enunciated new principles in science; some pushed forward the bounds of knowledge by new syntheses of our records of the past. Each lecture, now offered in essay form, provides an exciting, personal report of adventures on the frontier.

This idea, ingenious in its simplicity, is that of Daniel J. Boorstin, former Director of the National Museum of History and Technology and currently Senior Historian here. Those who heard the lectures enjoyed seeing and feeling the very different personalities of a wide range of creative people, of pioneers of a special sort. Those who now read the collected volume of essays have the opportunity to feel the multiplied effect of many accounts of individual perseverance and imagination.

Three series of lectures are included here, each dedicated to a single theme or responding to a single question. The lecturers in each series answer the thematic question from a different field and viewpoint. The first series contemplates the relationships of "Technology and the Frontiers of Knowledge," the second probes the questions of "Creativity and Collaboration," and the third relates experiences of "The Modern Explorers."

"Technology and the Frontiers of Knowledge" opens one of the broadest areas of inquiry but carefully follows the thread through history, into the present, and beyond the frontier of our own time into the future. All the authors give some thought to the over-all problem of man's relationships with his technology; each concentrates on a more limited aspect of the problem. The richness of the relationships in literature is effectively pursued by Saul Bellow, one of the most impressive novelists of our time. Technology and society has been the subject of books and symposia; Daniel Bell, one of our most influential sociologists, offers his own highly individualistic reactions. Edmundo O'Gorman, known for his unique interpretations of history, seeks to explore some of the meanings of technology in history. Sir Peter Medawar, an honored biologist, looks at one of the most serious problems—man's own extensions of the evolutionary process. Arthur C. Clarke, an explorer of the future through science fiction, is primarily concerned here with the historical record of converting unknowable things into the known through new technology and new techniques.

This series gives a happy start to the volume, because it extends the realms and aspirations of the National Museum of History and Technology in a series of stimulating statements. Knowledge emerges as an expanding volume, the surface of which moves outward at differing and unpredictable rates. However, everywhere, at the border between the known and the unknown, one phenomenon is present: the more man learns, the more he discovers that he wants to know.

In the second series, "Creativity and Collaboration," five leading creative spirits, each from a different realm, confront the question of what the need to collaborate has meant to him, personally, in pursuing creative goals. These lecturers, then, are not scholars or students of episodes they know only from a study of the record. They are the actors; they are the participants themselves, and they comment as eyewitnesses.

Akio Morita, a leading Japanese businessman, founder of the Sony Corporation, records the creativity of the inventor but also of the engineer and the business entrepreneur. He seems to feel that collaboration can be creative and that the innovations he has pioneered must have a collaborative character. On the other hand, James D. Watson, known for his role in the discovery of the DNA molecule, does not favor large-scale collaboration and asserts that two persons may, indeed, be the largest effective group. Huw Wheldon, managing director of television for the British Broadcasting Corporation, similarly tilts against large-scale team efforts. He documents clearly some of the spectacular successes of the BBC which depended heavily upon the idiosyncratic ingenuity of a single individual.

The viewpoint of the artist is presented by Moshe Safdie, architect and city planner, designer of "Habitat." Safdie's message is a familiar theme in architecture but one that he has translated into wholly new results. He emphasizes the role of the architect as an interpreter of social patterns and needs, but equally reveals the integrating role of the lone artist. Caspar W. Weinberger, Secretary of Health, Education, and Welfare under Presidents Nixon and Ford, undertakes and succeeds in an unusually difficult task, the demonstration that budget making is both a creative art and a challenging field of experience. Inescapably a great collaboration, government budget making nevertheless requires individual creativity.

These authors confront the question directly, but they

could not, nor could anyone, give definitive answers. Clearly, the role of collaboration varies from period to period, from field to field, from individual to individual, and even from specific problem to specific problem. Collaboration sometimes thrives upon and stimulates creativity. At other times, it may well be the antagonist, as some of these authors firmly assert.

The final series, "The Modern Explorers," presents some firsthand accounts of explorations that have been possible only in this century. Sir Edmund Hillary represents the classic explorer who ventured into the earth's last unknown lands. In presenting his account, he documents the expected but always remarkable courage and resourcefulness that accompany the intrepid frontiersman, whether climbing Mount Everest or crossing the Antarctic Continent. Sir Fred Hoyle is an explorer of an altogether different sort. One of the most controversial astronomers of our day, a leading creative and imaginative explorer of the unknown realms of time and space, he here suggests a resolution of the long-conflicting "steady-state" and "big bang" theories of the origin of the universe in a comprehensive innovation that seeks to combine the two.

Willard F. Libby, another major scientific creator, explains how he succeeded so decisively in the construction of a wholly new tool capable of exploring the past, for which he received a Nobel prize—carbon 14 dating. The revolutionary technique of radiocarbon dating was theorized and dramatically proved by Libby, and has since been used for more and more explorations by anthropologists, archaeologists, geologists, and historians. Isaac Asimov, effective writer in many realms, is almost an inhabitant of those worlds of science fiction of which he is master. He has attempted the difficult task of drawing projections beyond our landing on the moon by showing a series of possible

Foreword

curves of what might follow. The exercise is as interesting as the books that have made him famous.

The strength of this volume is the combined strengths of the creative men who have written it. They write of their roles in advancing knowledge and they write here in a directed diffusion of their many achievements. The result is stimulating to the mind; we may hope that it will stimulate still further advances in knowledge.

BROOKE HINDLE
Director
National Museum of History and Technology
Washington, D.C.

CONTENTS

Contents

THE FRONTIERS
OF KNOWLEDGE

FIRST SERIES

TECHNOLOGY AND THE FRONTIERS OF KNOWLEDGE

1972–73

SAUL BELLOW

DANIEL BELL

EDMUNDO O'GORMAN

SIR PETER MEDAWAR

ARTHUR C. CLARKE

LITERATURE IN THE
AGE OF TECHNOLOGY

SAUL BELLOW, *a master of fiction and one of the nation's best-known novelists, is a professor of English at the University of Chicago and Chairman of its Committee on Social Thought.*

Bellow has written, since a start at writing given him by the Federal Writers' Project, eight novels, five plays, and numerous short stories and essays. His first novels were THE DANGLING MAN (*1944*) *and* THE VICTIM (*1947*). *For* THE ADVENTURES OF AUGIE MARCH (*1953*), *Bellow received a National Book Award. The original edition of* SEIZE THE DAY (*1957*) *includes his short stories and a one-act play. In 1959, Bellow published* HENDERSON THE RAIN KING. *Two more recent novels,* HERZOG (*1964*) *and* MR. SAMMLER'S PLANET (*1969*), *have brought him further National Book Awards and international prizes.* HUMBOLDT'S GIFT (*1975*) *is his eighth and newest novel.*

NINETEENTH-CENTURY writers disliked or dreaded science and technology. Edgar Allan Poe, discovering that scientific attitudes could be richly combined with fantasy, created science fiction. A Shelley experimented romantically with chemicals; a Balzac thought himself a natural historian or social zoologist; but, for the most part, science, engineering, technology horrified writers. To mechanical energy and industrial enterprise, mass production, they opposed feeling, passion, "true work," artisanship, well-made things. They turned to nature, they specialized in the spirit, they valued love and death more than technical enterprise. Writers then preferred, and still do prefer, the primitive, exotic, and irregular. These romantic attitudes produced masterpieces of literature and painting. They produced also certain cultural platitudes. The platitude of dehumanizing mechanization formed on the one side. Equally platitudinous, the vision of a new age of positive science and of rational miracles, of progress, a progress that made art as obsolete as religion, filled up the other horizon. The platitude of a dehumanized technology gives us, today, novels whose characters are drug-using noble savages, beautiful, mythical primitives who fish in the waters dammed up by mighty nuclear installations. And, as power-minded theoreticians see it, the struggle between old art and new technology has ended in the triumph of technology. The following statement, and it is a typical one, is made by Mr. Arthur C. Clarke in a book called *Report on Planet Three:*

4

Literature in the Age of Technology

It has often been suggested that art is a compensation for the deficiencies of the real world; as our knowledge, our power and above all our maturity increase, we will have less and less need for it. If this is true, the ultra-intelligent machine would have no use for it at all.

Even if art turns out to be a dead end, there still remains science. . . .

This statement by a spokesman of the "victorious" party is for several reasons extraordinarily silly. First, it assumes that art belongs to the childhood of mankind, and that science is identical with maturity. Second, it thinks art is born in weakness and fear. Third, in its happy worship of "ultra-intelligent" machines, it expresses a marvelous confidence in the ability of such machines to overcome all the deficiencies of the real world. Such optimistic rationalism is charming, in a way. Put it into rhyming verses and it may sound a lot like Edgar A. Guest. Edgar felt about capitalism and self-reliance precisely as Arthur C. Clarke feels about the supertechnological future. They share a certain expansiveness, the intoxication of the winner, the confidence of the great simplifier. Mr. Clarke says in effect, "Don't worry, dear pals, if art is a dead end, we still have science. Pretty soon we won't need Homer and Shakespeare, Monteverdi and Mozart. Thinking machines will give us all the wisdom and joy we want, in our maturity."

I have chosen a different sort of theorist to put the question from another angle. In the *Atlantic* of July 1972, Mr. Theodore Roszak takes issue with Robert J. Good, a professor of chemical engineering. In a letter to the magazine Professor Good, of the party of science and technology, says that it is sad to see modern intellectuals "cutting off their own roots in rationality." Mr. Roszak tries to deal gently with the professor. He says with a pious tremor,

The Frontiers of Knowledge

What can one do, even in radical dissent, but handle with affectionate care so noble and formidable a tradition within our culture—even knowing it is a tragic error and the death of the soul? It is not primarily science I pit myself against in what I write. Rather, the wound I seek to heal is that of psychic alienation: the invidious segregation of humanity from the natural continuum, the divorce of visionary energy from intellect and action. What Professor Good disparages as the irrational (Lord! must I, too, centuries after Blake, repeat the lesson in this limp prose?) is a grand spectrum of human potentialities. When rationality is cut away from that spectrum, then the life of Reason becomes that mad rationality which insists that only what is impersonal and empirical, objective and quantifiable is real—*scientifically* real. Believe that and you are not far from tabulating the tragedies of our existence by way of body counts and megadeaths and chemical imbalances within our neural circuitry.

Of course such issues cannot be discussed without invoking Vietnam, or whatever is the most monstrous topic of the day. For ideologists in all fields the political question is always hugely, repulsively, squatting behind a paper screen.

But when we have cited the argument that ultra-intelligent technology has no need of art, and the counterargument that creativity is needed to heal psychic alienation and keep us from criminal wars, we have not altogether exhausted our alternatives. There is a third alternative, which has nothing much to do with compensation for the deficiencies of the world or with society's health. This alternative holds that man is an artist, and that art is a name for something always done by human beings. The technological present may be inhospitable to this sort of doing, but art can no more be taken from humankind than faces and hands. The giving of weight to the particular, and the

6

tendency to invest the particular with resonant meanings, cannot be driven out by the other tendency: to insist on the finitude of the finite and to divest it of awe and beauty.

Is the film *2001*, with a sinister computer that speaks in a homosexual voice, a forerunner of the new maturity? Will this sort of drama replace *Othello*, in which an immature husband murders his wife—I call Othello immature, because I have been told by a famous progressive psychiatrist that in future ages, with sexual jealousy gone, Othello will be incomprehensible. Scientism dearly loves to speak of the childish past, the grave future. When I hear people invoking the maturity of future generations, I think of a conversation in the *Anti-Memoirs* between Malraux and a parish priest who joined the Resistance and later died fighting the Germans. Malraux asked this man,

"How long have you been hearing confessions?"
"About fifteen years."
"What has confession taught you about men?"
". . . First of all, people are much more unhappy than one thinks . . . and then . . . the fundamental fact is that *there's no such thing as a grown-up* person. . . ."

Anyway, romantics hold that it is very dangerous to sanity to deny the child within us, while scientism says that technological progress is about to carry us for the first time into the adult stage. On both sides, intellectuals take positions on what art can or cannot add to human happiness. In most discussions the accent falls on health, welfare, progress, or politics—anything but art. About art itself, most intellectuals know and care little.

Malraux begins his *Anti-Memoirs* with the wise chaplain of the Maquis. He then goes immediately into the subject of memoirs and confessions and discusses the "theatrical self-image" in autobiography. How to reduce to a minimum the theatrical side of one's nature is a matter of great concern to him.

Once, man was sought in the great deeds of great men; then he was sought in the secret actions of individuals (a change encouraged by the fact that great deeds were often violent, and the newspaper has made violence commonplace).

Malraux then concludes,

. . . the confessions of the most provocative memorialist seem puerile by comparison with the monsters conjured up by psychoanalytic exploration, even to those who contest its conclusions. The analyst's couch reveals far more about the secrets of the human heart, and more startlingly, too. We are less astonished by Stavrogin's confession [in Dostoevsky's *The Possessed*] than by Freud's *Man with the Rats:* genius is its only justification.

The genius, I take it, belongs to Dostoevsky. I am not quite sure what Malraux means, but I think he is saying that what a psychoanalyst learns about the human heart is far deeper and more curious than anything the greatest novelists can reveal. Perhaps he hints even that the madman is more profoundly creative in his rat-imaginings or grotesque fantasies than the writer, who can only compensate us for the ordinariness of his "vile secrets" or "frightful memories" by the power of his mind ("genius is [his] only justification"). But even if we do not push the matter so far, we can legitimately take Malraux, a novelist, to be saying that in the field of facts the writer is "puerile." He cannot compete in that field with the clinical expert.

So novelists once gave "information" to the public. But when people now really want to know something, they turn to the expert. Universities and research institutes produce masses of experts, and governments license them. The dazzle of expertise blinds the unsure, the dependent, and the wretched. It is not the novelist alone who has lost ground. Expertise has made all opinion shaky, and even powerful men are reluctant to trust their own judgment. In totalitarian countries, where facts are suppressed, writers of exceptional courage still tell the truth in the old way. (Why was it not a Soviet expert who told the world about the Gulag Archipelago?) But in the Free World, novelists are peculiarly inhibited.

Artists were great and highly visible monuments in the nineteenth century. The public listened deferentially to its Victor Hugos and its Tolstoys. But Shaw and Wells were the last of these prestigious literary spokesmen. In the postwar period, only Bertrand Russell and Jean-Paul Sartre appeared before the world in this role, and even if these two had been more consistent and sensible they would not have affected the public as their greater predecessors had done. The era of the writer as public saga and as dependable informant has ended.

A single standard has been set for novelists and for experts: the fact standard. The result of this strict accountability has been to narrow the scope of the novel, to make the novelist doubt his own powers and the right of his imagination to range over the entire world. The authority of the imagination has declined. This has had two remarkable results. Earlier in the century, certain writers rejected the older novel with its more modest objectives. The Dickens sort of novel—*Great Expectations*, say—was replaced by the more comprehensive novel, nothing less than an aesthetic project for encompassing the whole world. Books such as Joyce's *Ulysses* and Proust's *Remembrance of Things Past* do not draw the real world so much as they replace it by aesthetic fiat. Without perhaps intending to be one, Joyce was, in effect, an aesthetic dictator. The century needed a book? He provided one. It was a book that made other books unnecessary. It had taken about twenty years to write *Ulysses* and *Finnegans Wake*, and it should take just as long to read them. If you devoted two decades to Joyce, you'd have no time for other writers. Anyway, your need for literature would be fully satisfied. This was one result of the weakening of the authority of writers and of the power of the literary imagination to command attention—overassertiveness. A more recent result has been surrender. Writers have capitulated to fact, to events and reportage, to politics and demagogy. Modern art has tried to create power for itself on arbitrary terms and has also pursued and worshiped power in its public forms.

Until recent times, the artist's dream sphere was distinctly separate from the practical or mechanical realm of the technician. But in the twentieth century, as Paul Valéry recognized, a change occurred. He wrote in an essay:

The fabulous is an article of trade. The manufacture of machines to work miracles provides a living to thousands of people. But the artist has had no share in producing these wonders. They are the work of science and capital. The bourgeois has invested his money in phantoms and is speculating on the downfall of common sense.

Yes, technology is the product of science and capital, and of specialization and the division of labor. It is a triumph of the accurate power of innumerable brains and wills acting in unison to produce a machine or a commodity. These many wills constitute a fictive superself astonishingly effective in converting dreams into machines. Literature, by contrast, is produced by the single individual, concerns itself with individuals, and is read by separate persons. And the single individual, the unit of vital being, of nerve and brain, who judges or knows, is happy or mourns, actually lives and actually dies, is unfavorably compared with that fictive superself which, acting in unison and according to plan, produces jet planes, atomic reactors, computers, rockets, and other modern technological wonders. Glamorous, victorious technology is sometimes considered to have discredited all former ideas of the single self.

To theorists of the new, a thing is genuine only if it manifests the new. Valéry in the essay *Remarks on Progress*, from which I have just quoted, illustrates this attitude remarkably well. He says,

Men are doubtless developing the habit of considering all knowledge as transitional and every stage of their industry and their relations as provisional. This is new.

And again,

Suppose that the enormous transformation which we are living through and which is changing us, continues to develop, finally altering whatever customs are left and making a very different adaptation of our needs to our means; the new era will soon produce men who are no longer attached to the past by any habit of mind. For them history will be nothing but strange, incomprehensible tales; there will be nothing in their time that was ever seen before—nothing from the past will survive into their present. Everything in man that is not purely physiological will be altered, for our ambitions, our politics, our wars, our manners, our arts are now in a phase of a quick change; they depend more and more on the positive sciences and hence less and less on what used to be. *New facts* tend to take on the importance that once belonged to tradition and *historical facts*.

This is the quintessence of the tradition of the new. By attaching itself to technology, "newness" achieves a result longed for by those thinkers of the previous century who were oppressed by historical consciousness. Karl Marx felt in history the tradition of all the dead generations weighing like a nightmare on the brain of the living. Nietzsche speaks movingly of the tyranny of "it was," and Joyce's Stephen Dedalus also defines history as a nightmare from which we are trying to awaken. The vision of freedom without conditions, a state of perfect and lucid consciousness into which we are released by technological magic from all inertias, is a sort of romance, really, a French intellectual's paradise. But Valéry does not neglect the painful side of this vision:

. . . one of the surest and cruelest effects of progress is to add a further pain to death, a pain increasing of itself as the revolution in customs and ideas becomes more marked and rapid. It is not enough to perish; one has to become unintelligible, almost ridiculous; and even a Racine or a Bossuet must take his place alongside those bizarre figures, striped and tattooed, exposed to passing smiles, and somewhat frightening, standing in rows in the galleries and gradually blending with the stuffed specimens of the animal kingdom. . . .

So, at the height of technological achievement there blazes the menace of obsolescence. The museum, worse than the grave because it humiliates us by making us dodos, waits in judgment on our ambitions and vanities. Of course, no one wants to suffer the double doom of obsolescence—to be dead and also to be a fossil. Everyone wants to be the friend and colleague of history. And consciously or not, intellectuals try hard to be what Hegel called Historical Men or World-Historical Individuals, those persons through whom truth operates and who have an insight into the requirement of the time, who divine what is ripe for development, the nascent principle, the next necessary thing. They may denounce the nightmare past, but they have also an immortal craving to be in the line of succession, and to prove themselves to be historically necessary. It is these people, lovers of the new, who derive from technological progress a special contempt for the obsolete. The enemies of pastness, even though they tell us that we will depend more and more on the positive sciences and hence less and less on what used to be, insofar as they seek the next necessary development, make their own kind of historical judgment. Intellectuals, when they sense the cruel effects of technological progress, try not only to escape oblivion themselves by association with the next necessary thing, but also to impose oblivion on others—on those writers who fail to recognize that the human condition has been, or will be, completely transformed by science and the revolution of customs, and who, in the old-fashioned solitude of old-fashioned rooms, continue to consider the destinies of old-fashioned individuals and follow their old-fashioned trade (a home industry of the seventeenth century), unaware that the World Spirit has abandoned them.

Unlike Huxley's *Brave New World* or George Orwell's *Nineteen Eighty-four*, Joyce's *Ulysses* is not directly concerned with technology. It remains nevertheless the twentieth century's most modern novel—it is *the* account of human life in an age of artifacts. Things in *Ulysses* are not nature's things. Here the material world is wholly man's world, and all its objects are human inventions. It is made in the image of the conscious mind. Nature governs physiologically, and of course the unconscious remains nature's stronghold, but the external world is a world of ideas made concrete. Between these two powers, nature within, artifacts without, the life of Mr. Leopold Bloom is comically divided. The time is 1904. No one in Dublin has seen Mr. Arthur C. Clarke's ultra-intelligent machine even in a dream, but the age of technology has begun and *Ulysses* is literature's outstanding response to it.

Now, what is *Ulysses?* In *Ulysses* two men, Dedalus and Bloom, wander about the city of Dublin on a June day.

Mrs. Bloom, a singer, lies in bed, reading, misbehaving, musing and remembering. But nothing that can be thought or said about human beings is left out of this account of two pedestrians and an adulteress. No zoologist could be more explicit or complete than Joyce. Mr. Bloom, first thing in the morning, brews the tea, gives milk to the cat, goes to the pork butcher to get meat for his breakfast, carries a tray up to his wife, eats a slightly scorched kidney, goes out to the privy with his newspaper, relieves himself while reading a prize-winning story, wipes his bottom with a piece of the same paper, and then goes out to the funeral of Paddy Dignam. Matters could not be more real.

Now, realism in literature is a convention, and this convention postulates that human beings are not what everyone for long centuries conceived them to be. They are something different, and they live in a disenchanted world that exists for no particular purpose that science can show. Still, people continue to try to lead a human life. And this is rather quaint, because man is not the comparatively distinguished creature he thinks himself to be. The commonness of common life was a great burden to nineteenth-century writers. The best of them tried to salvage something from the new set of appalling facts. Coleridge tells us how Wordsworth intended to redeem the everyday by purging the mind of "the lethargy of custom" and showing us the beauty and power of what we call commonplace. He did not do this to the satisfaction of his successors. English writers of the second half of the century were much more impressed by the weight given to the evil component of the commonplace by their French contemporaries. A novelist like Flaubert saw nothing in the banal average that did not disgust him. But art, virtuosity, language, the famous objectivity, these, after painful struggle, would make commonplace reality yield gold.

Joyce, a Flaubertian to begin with, gives in *Ulysses* the novel's fullest account of human life—within this realistic convention. As he sees it, the material world is now entirely human. Everything about us—clothing, beds, tableware, streets, privies, newspapers, language, thought—is man-made. All artifacts originate in thought. They are thoughts practically extended into matter.

Joyce is the complete naturalist, the artist-zoologist, the poet-ethnographer. His account of Bloom's life includes everything. Everything seems to demand inclusion. No trivialities or absurdities are omitted. Old bourgeois reticences are overrun zigzag. For what, after all, is the important information? No one knows. Anything at all may be important. Freud taught, in *The Psychopathology of Everyday Life*, that the unconscious did not distinguish between major and minor matters, as conscious judgment did, and that the junk of the psyche had the deepest things to tell us. Joyce is the greatest psychic junkman of our age, after Freud. For the last of the facts may be the first. Thus we know the lining of Bloom's hat, and the contents of his pockets; one knows his genitals and his guts, and we are thoroughly familiar with Molly, too, how she feels before her period, and how she smells. And with so much knowledge, we are close to chaos. For what are we to do with such a burden of information? *Ulysses* is a comedy of information. Leopold Bloom lies submerged in an ocean of random facts: textbook tags, items of news, bits of history, slogans, clichés, ditties, operatic arias, saws and jokes, scraps of popular science, and a great volume of superstitions, fantasies, technical accounts of the Dublin water supply, observations about hanged men, recollections of copulating dogs. The debris of learning, epic, faith, and enlightenment pour over him. In this circumambient ocean, he seems at times to dwell like a coelenterate or a sponge. The man-made world begins, like the physical world, to suggest infinity. The mind is endangered by the multitude of accounts that it can give of all matters. It is threatened with inanity or disintegration.

William James believed that not even the toughest of tough minds could bear to know everything that happened in a single city on a given day. Not everything. No one could endure it. It is probably one of the functions of the nervous system to screen us and to preserve us from disintegrating in the sea of facts. We ourselves, however, seek out this danger, the Faustian dream of omniscience lives on. To Joyce this Faustian omniscience is a deliciously funny theme. —I assume that Joyce knew Flaubert's last novel, of two elderly cranks, Bouvard and Pécuchet, who in retirement try to investigate every branch of knowledge.

At all events, Bloom's mind is assailed and drowned by facts. He appears to acknowledge a sort of responsibility to these facts, and he goes about Dublin doing his facts. This suggests that our scientific, industrial, technical, urban world has a life of its own and that it borrows our minds and souls for its own purposes. In this sense, civilization lives upon Bloom. His mind is overcome by its data. He is the bearer, the servant, the slave of involuntary or random cognitions. But he is also the poet of distractions. If Bloom were only the *homme moyen sensuel*, or everyman, nothing but the sort of person realism describes as "ordinary," he would not be the Bloom we adore. The truth is that Bloom is a wit, a comedian. In the depths of his passivity, Bloom resists. He is said in Dublin to be "something of an artist." To be an artist in the ocean of modern information is certainly no blessing. The artist has less power to resist the facts than other men. He is obliged to note the particulars. One may even say that he is condemned to see them. In the cemetery, Bloom can't help seeing the gravedigger's spade and noting that it is "blueglancing." He is therefore receptively, artistically, painfully immersed in his mental ocean. The fact that he is "something of an artist" aggravates the problems of information. He seeks relief in digression, in evasion, and in wit.

Why is the diversity of data so dazzling and powerful in *Ulysses?* The data are potent because the story itself is negligible. *Ulysses,* as Gertrude Stein once said, is not a "what-happens-next?" sort of book. A "what-happens-next?" story would, like a nervous system, screen out distractions and maintain order.

It is the absence of a story that makes Bloom what he is. By injecting him with purpose, a story would put the world in order and concentrate his mind. But perhaps Bloom's mind is better not ordered. Why should he, the son of a suicide, the father who mourns a dead child, a cuckold, and a Jew in Catholic Dublin, desire moral and intellectual clarity? If his mind were clear, he would be another man entirely. No, the plan of Bloom's life is to be planless. He palpitates among the phenomena and moves vaguely toward resolution. Oh, he gets there, but *there* is a region, not a point. At one of the low hours of his day he thinks, "Nothing is anything." He feels his servitude to the conditions of being. When there is no story, those conditions have it all their own way and one is delivered to despair. The artifact civilization, Joyce seems to tell us, atrophies the will. The stream of consciousness flows full and wide through the will-less. The romantic heroes of powerful will, the Rastignacs and the Raskolnikovs, are gone. The truth of the present day is in the little Blooms, whose wills offer no hindrance to the stream of consciousness. And this stream has no stories. It has themes. Bloom does not, however, disintegrate in the thematic flow. Total examination of a single human being discloses a most extraordinary entity, a comic subject, a Bloom. Through him we begin to see that anything can be something. But the burden of being a Bloom is nevertheless frightful. It is not clear exactly how Joyce would like us to see the Bloom problem. Long passages of *Ulysses* are bound together by slurs (in the musical sense of the word) of ambiguous laughter. It does, however, appear that Joyce expected the individual who has gone beyond the fictions and postures of "individuality" (romantic will, etc.) to be sustained by suprapersonal powers of myth. Myth, rising from the unconscious, is superior to mere "story," but myth will not come near while ordinary, trivial ideas of self remain. The powers of myth can be raised up only when the discredited pretensions of selfhood are surrendered. Therefore consciousness must abase itself, and every hidden thing must be exhumed. Hence Bloom's moments in the privy, his corpse fantasies at Paddy's funeral, his ejaculation as he watches crippled Gerty, his masochistic hallucinations in Nighttown. The old dignities must take a terrific beating in this new version of "the first shall be last and the last first."

How is the power of the modern age to be answered? By an equal and opposite power within us, tapped and interpreted by a man of genius. Joyce performs the part of the modern genius to perfection. This sort of genius, as I see it, comes from the mass without external advantages. Everything he needs is within him. By interpreting his own dreams, he creates a scientific system. By musical spells extracted from his own personality, he hypnotizes the world with his Siegfrieds and Wotans. He is, as Collingwood calls him, the "mystagogue leading his audience . . . along the dark and difficult places of his own mind . . . the great man who (as Hegel says) imposes upon the world the task of understanding him." This task of understanding has certainly been imposed by Joyce. These men of genius take you in their embrace and propose to be everything to you. They are your *Heimat*, your church. You need no other music than theirs, no other ideas, no other analysis of dreams, no other manna. They are indeed stirring and charming. Their charms are hard to get away from. Theirs are the voices under which other voices sink. Once initiated into their mysteries, we do not easily free ourselves.

It is now seventy years since Bloom walked the streets of Dublin. In those seventy years the noise of life has increased a thousandfold. And already at the beginning of the nineteenth century Wordsworth was alarmed by the increase of distractions. The world was too much with us. The clatter of machinery, business, the roar of revolution would damage the inmost part of the mind and make poetry impossible. A century after Wordsworth, ingenious Joyce proposed to convert the threat itself into literature.

What you feel when you read *Ulysses* today is the extent to which a modern society imposes itself upon everyone. The common man, who, in the past, knew little about the great world, now stands in the middle of it. At least he thinks he does, as reader, hearer, citizen, voter, and judge of all public questions. His imagination has been formed to make him think himself in the center. The all-important story appears to belong to society itself. Real interest is monopolized by collective achievements and public events, by the fate of mankind, by a kind of "politics." The voluminous Sunday *Times* is put into our hands together with *Time* and *Newsweek*, while images from television flash before us. This is the week's record of everything of substance relating to the human species. It is about us, our hope for survival, our common destiny. Is it, now? Does this really speak to my condition? Is this mankind, is it *me*, heart, soul, and destiny? The nominally central individual studying the record does not in fact feel central. On the contrary, he feels peculiarly contentless in his public aspect, lacking in substance and without a proper story. A proper story would express his intuition that his own existence is peculiarly significant. The sense that his existence is significant haunts him. He can prove nothing. But the business of art is with this sense, precisely.

Though the sun shines sweetly, the modern mind knows that there are devilish processes of nuclear fusion and staggering explosions in the heavens. So, as mild Bloom goes down the streets, we are aware of a formidable intellect that follows him as he buys a cake of soap. The modernists are learned intellectuals—Viconians like Joyce, Freudians, Marxians, Bergsonians, et cetera. A technological society produces mental artists and an intensely intellectual literary culture. "Most modern masterpieces are critical masterpieces," writes Harold Rosenberg, who thinks that this is as it should be. "Joyce's writing is a criticism of literature, Pound's poetry a criticism of poetry, Picasso's painting a criticism of painting. Modern art also criticizes the existing culture. . . . One keeps hoping that the decline in excellence of people and things is an effect of transition. All we have on the positive side is the individual's capacity for resistance. Resistance and criticism." To hope that the decline in the excellence of people and things (the last an effect of technology) is an effect of transition, shows Mr. Rosenberg's heart to be in the right place. But the emphasis on criticism shows something else, namely a claim for intellectual priority. Art is something that must satisfy the requirements of intellectuals. It must interest intellectuals by being, in the right sense, critical of the existing culture. The fact is that modern art has tried very hard to please its intellectual judges. Intellectual judgment in the twentieth century very much resembles aristocratic taste in the eighteenth—in the sense, only, that artists in both centuries acknowledge its importance. Art in the twentieth century is more greatly appreciated if it is directly translatable into intellectual interests, if it stimulates ideas, if it lends itself to discourse. Because intellectuals do not like to suspend themselves in works of the imagination. They want to talk. Thus they make theology and philosophy out of literature. They make psychological theory. They make politics. Art is one of the principal supports of this social class.

Gide's *The Counterfeiters* is a cultural product as well as a novel; *The Red and the Black* is no such thing. *The Magic Mountain* belongs to intellectual history, a category that does not exist for that excellent book *Little Dorrit*. It never occurred to Dickens to run over into cultural criticism or to be Carlyle and Mill as well as Charles Dickens. But, in the twentieth century, writers are often educated men as well as creators, and in some the education prevails over the creation. There are reasons for this. A burden of "understanding" has been laid upon us by this revolutionary century. What I am trying to indicate is that cultural style is not to be confused with genuine understanding. At the moment, such understanding has few representatives, while cultural style seems to have hundreds of thousands.

One of the problems of literature in this age of technology is the problem of those who preside over literary problems, of specialists, scholars, historians, and teachers. Modern writers, themselves more "intellectual" than those of a century ago, face a public formed, educated, and dominated by professors, "humanists," "anti-humanists," by psychologists and psychotherapists, by the professional custodians of culture, and by ideologists and shapers of the future. This critical public has a thousand *important* (i.e., political and social) questions to answer. It is irritably fastidious. It asks,

"To whom should we listen? Who, if anyone, can be read? What is, or can be, really interesting to a modern cultivated intelligence?"

Such questions can only be answered with sadness and sighs. To prove that I am not exaggerating, I shall quote briefly from an essay by Lionel Trilling, in *Commentary*, November 1972, "Authenticity and the Modern Unconscious." In this essay Professor Trilling observes that in this day and age, things being what they are, novels can no longer be Authentic or appeal to Authentic readers. "It is the exceptional novelist today who would say to himself, as Henry James did, that he 'loved the story as story,' by which James meant the story apart from any indeational intention it might have, simply as, like any primitive tale, it brings into play what James called 'the blessed faculty of wonder.' Already in James's day, narration as a means by which the reader was held spellbound, as the old phrase put it, had come under suspicion. And the dubiety grew to the point where Walter Benjamin could say some three decades ago that the art of storytelling was moribund."

Here one cries out, "Wait! Who is this Benjamin! Why does it matter what *he* said?" But intellectuals do refer to one another to strengthen their arguments. It turns out that the late Walter Benjamin objected to storytellers because they had an orientation toward practical interests. Stories, Trilling quotes Benjamin as saying, are likely to contain "something useful." Here what I have called "cultural style" begins to show itself. Modern literary culture, which prides itself on being radical, dissenting, free, has its own orthodoxy. Don't we know how it views the Bourgeois, the Child, the Family, Technology, the Artist, the Useful? We do indeed. The idea of usefulness, Baudelaire said, nauseated him. And there is the foundation of your orthodoxy. Storytellers, Benjamin objects, have "counsel to give" and this giving of counsel has "an old-fashioned ring." Professor Trilling then continues, It is "inauthentic for our time— there is something inauthentic for our time in being held spellbound, momentarily forgetful of oneself, concerned with the fate of a person who is not oneself. . . . By what right, we are now inclined to ask, does the narrator exercise authority over that other person, let alone over the reader, and arrange the confusion between the two, and presume to have counsel to give?"

If there is something old-fashioned and inauthentic for our time in being held spellbound, then Homer and Dostoevsky, whose works hold us spellbound, are inauthentic. The point seems to be that the modern condition is killing certain human activities (arts) once highly valued. For an Authentic modern man, living in a modern technological society, naïve self-surrender is impossible. Apparently the question is partly one of authority. "By what right does the narrator" presume to invade our minds, deliberately confuse us, and give counsel? I don't see what good it does to make a political question of this. By what right do our parents conceive us, or we our children? By what right does society teach us a language or give us a culture? If Authentic man had no words, he would be unable to express his longing to be so virginal.

But this, I realize, is not quite fair. Professor Trilling wishes to leave the surface of life with its stories and descend into the depths in search of truth and maturity, becoming one of Aristotle's great-souled men. Thus Professor Trilling seems to agree, with Malraux's priest in the Resistance, that there is no such thing as a grown-up person. His position is also close to that of Mr. Arthur C. Clarke, who suggests that art is a compensation for the deficiencies of the real world and that "as our knowledge, our power and above all our *maturity* increase, we will have less and less need for it. . . . the ultra-intelligent machine would have no use for it at all."

So Professor Trilling, moving toward "scientific truth," reports that we can no longer be held spellbound, and Mr. Clarke tells us that we can be redeemed by technology from the childish need for art. Perhaps a modest, fair statement of the case is that human beings have always told stories to one another. By what *right* have they done this, and on what authority? Well, on none, really. They have obeyed the impulse to tell and the desire to hear. Science and technology are not likely to remove this narrative and spellbinding oddity from the soul. The present age has a certain rationalizing restlessness or cognitive irritability; a participatory delirium that makes the arresting powers of any work of art intolerable. The desire to read is itself spoiled by "cultural interests" and by a frantic desire to associate everything with something else and to convert works of art into subjects of discourse. Technology has weakened certain points of rest. Wedding guests and ancient mariners both are deafened by the terrific blaring of the technological band.

In a charming and strange book, the prerevolutionary Russian writer V. V. Rozanov argues against repressive puritanism in words that can be applied more widely and are relevant to the subject of this paper. He writes:

A million years passed before my soul was let out into the world to enjoy it; and how can I say to her, "Don't forget yourself, darling, but enjoy yourself in a responsible fashion."

No, I say to her: "Enjoy yourself, darling, have a good time, my lovely one, enjoy yourself my precious, enjoy yourself in any way you please. And toward evening you will go to God."

For my life is my day, and it is my day. and not Socrates' or Spinoza's.

Thus, to the queen or tramp who is his soul, Rozanov speaks with an erotic-religious aim of some sort. But we can adapt this to our own purpose, saying, "A million years passed before my soul was let out into the technological world. That world was filled with ultra-intelligent machines, but the soul, after all, was a soul, and it had waited a million years for its turn and did not intend to be cheated of its birthright by a lot of mere gimmicks. It had come from the far reaches of the universe and it was interested but not overawed."

TECHNOLOGY, NATURE,
AND SOCIETY

DANIEL BELL, *professor of sociology at Harvard University, is best known for his wide-ranging and original essays of social criticism. He is the author, among other works, of* THE END OF IDEOLOGY *(1960),* THE REFORMING OF GENERAL EDUCATION *(1965), and* THE COMING OF POST-INDUSTRIAL SOCIETY *(1973).*

He was vice-president (for the social sciences) of the American Academy of Arts and Sciences, and chaired its Commission on the Year 2000. Before coming to Harvard in 1969, Bell had served on the faculties of the University of Chicago and Columbia University, and was a Fellow at the Center for Advanced Study in the Behavioral Sciences at Stanford, California.

Bell was cofounder and coeditor of The Public Interest *and is now chairman of its publications committee. He has been on the editorial board of* The American Scholar *and now serves as a member of the Board of Editors of* Daedalus.

During the Johnson administration, Mr. Bell was a member of the President's Commission on Technology, Automation and Economic Progress (1964–66), and served as cochairman of the Panel on Social Indicators of the Department of Health, Education, and Welfare (1966–68).

27

I

THE TERMS of the will of James Smithson, as you know, bequeathed the whole of his property to the United States of America, "to found at Washington, under the name of the Smithsonian Institution, an Establishment for the increase and diffusion of knowledge among men." Though the bequest, in one sense, was clear, the effort to implement it led for several decades to many confusions and debates. What is knowledge, and how does one increase it or diffuse it? Some individuals wanted to create a national university, others a museum, still others a library, and others still a national laboratory, an agricultural experiment station, or, with John Quincy Adams, a national observatory. Today we have all these except a national university—though some local patriots might consider my home on the Charles such an institution. And certainly, under Mr. Dillon Ripley, the Smithsonian has become "an Establishment."

But if in later years buildings were built and institutions established, the more vexing question, of what knowledge should be increased and promoted, which bedeviled the regents of the Institution, still remains. In the mid-nineteenth century the "promotion of abstract science," as Joseph Henry, the first head of the Institution, put it, dominated the activities of the Smithsonian. But Mr. Henry soon found himself under attack from all sides. There were those like Alexander Dallas Bache, who said that ". . . a promiscuous assembly of those who call themselves men of science would only end in disgrace." Under the new conditions of scientific specialization, he declared, the universal savant was obsolete; the differentiation of scientists from amateurs demanded the material support only of professional research scientists. On the other hand, Horace Greeley, in the New York *Tribune*, accused Mr. Henry of converting the Smithsonian into "a lying-in hospital for a little knot of scientific valetudinarians." The question of what kind of science, theoretical or applied, continues to be refought.

A different, equally familiar issue was the one between men of science and men of letters. Ethics and philosophy, said Rufus Choate of Massachusetts, were as vital as soil chemistry and a knowledge of noxious weeds, and in the debate in the House of Representatives Choate's protégé, Charles W. Upham, representing the men of letters, declared: ". . . vindicate art, taste, learning, genius, mind, history, ethnology, morals—against sciologists, chemists & catchers of extinct skunks."[1]

In the unhappy further differentiation of the world since then, I appear here neither as a man of science nor as a man of letters. Sciologists (the bearers of superficial learning) have become crossed with logomachs (those who contend wearily about words) to create sociologists, that hybrid with a Latin foreword and a Greek root, symbolizing the third culture which has diffused so prodigiously throughout the modern world.

Yet as an intellectual hybrid my provenance may not be amiss. For my theme this evening is the redesign of the intellectual cosmos, the hybrid paths it has taken, and the necessary and hybrid forms it may take. With Mr. Upham's charge in mind, I am prepared to vindicate all his categories, except extinct skunks.

II

If we ask what uniquely marks off the contemporary world from the past, it is the power to transform nature. We define our time by technology. And until recently we have taken material power as the singular measure of the advance of civilization.

The philosophical justification of this view was laid down a hundred or more years ago by Marx. Man has needs which can only be satisfied by transforming nature, but in transforming nature he transforms himself: as man's powers expand he gains a new consciousness and new needs —technological, psychological, and spiritual—which serve, further, to stimulate man's activity and the search for new powers. Man, thus, is defined not by nature but by history. And history is the record of the successive plateaus of man's powers.[2]

But if it is, as Marx states in *Capital*, that in changing his external environment man changes his own nature, then human nature in ancient Greece must have been significantly different from human nature under modern capitalism, where needs, wants, and powers are so largely different. And if this is so, how is it possible, as Sidney Hook asks, to understand past historical experience in the same way we understand our own, since understanding presupposes an invariant pattern? This is a problem which confronts not only historical materialism but all philosophies of history.[3]

Marx only once, to my knowledge, in a fragment written in 1857, sought to wrestle with this conundrum; and his answer is extraordinarily revealing:

Technology, Nature, and Society

It is a well-known fact that Greek mythology was not only the arsenal of Greek art but also the very ground from which it had sprung. Is the view of nature and social relations which shaped Greek imagination and Greek [art] possible in the age of automatic machinery, and railways and locomotives, and electric telegraphs? Where does Vulcan come in as against Roberts & Co.; Jupiter as against the lightning rod; and Hermes as against the Crédit Mobilier? All mythology masters and dominates and shapes the forces of nature in and through imagination; hence it disappears as soon as man gains mastery over the forces of nature. . . . Is Achilles possible side by side with powder and lead? Or is the Iliad at all compatible with printing press and steam press? Do not singing and reciting and the muses necessarily go out of existence with the appearance of the printer's bar, and do they not, therefore, disappear with the prerequisites of epic poetry?

But the difficulty is not in grasping the idea that Greek art and epos are bound up with certain forms of social development. It rather lies in understanding why they still constitute with us a source of aesthetic enjoyment and in certain respects prevail as the standard and model beyond attainment.

The reason, Marx declares, is that such art is the *childhood* of the human race and carries with it all the charm, artlessness, and precocity of childhood, whose truths we sometimes seek to recapture and reproduce "on a higher plane." Why should "the social childhood of mankind, where it had obtained its most beautiful development, not exert an eternal charm as an age that will never return?"[4] That is why we appreciate the Greek spirit.

The answer is a lovely conceit. Yet one must know the sources of the argument to understand the consequences. For Marx, this view derived, in the first instance, from the conception of man as *homo faber*, the tool-making animal; the progressive expansion of man's ability to make tools is,

31

therefore, an index of man's powers. A second source of this view was Hegel, who divided history into epochs or ages, each a structurally interrelated whole and each defined by a unique spirit qualitatively different from each other. From Hegel, this view passed over into cultural history, with its periodization of the Greek, Roman, and Christian worlds, and Renaissance, Baroque, Rococo, and Modern styles. Sociologically, Hegel's idea is the basis of the Marxist view of history as successive slave, feudal, bourgeois, and socialist societies. Behind it all is a determinist idea of progress in human affairs, or a *marche générale* of human history, in which rationality in the Hegelian view, or the powers of production in the Marxist conception, are the immanent, driving forces of history that are obedient to a teleology in which anthropology, or a man-centered world, replaces theology, or a God-created world.

Today we know that, of the two views, that of *homo faber* is inadequate and that of the march of society and history is wrong. Man is not only *homo faber* but *homo pictor*, the symbol-producing creature, whose depictions of the world are not outmoded in linear history but persist and coexist in all their variety and multiplicity through the past and present, outside of "progressive" time. As for the nineteenth-century view of society, just as the mechanistic world view of nature has been destroyed by quantum physics, so the determinist theory of history has been contradicted by the twentieth-century clash of different time-bound societies.[5]

So we are back to our initial question: what marks off the present from the past, and how do we understand each other; how, for example, do we read the ancient Greeks, and how would they read us? The answer lies, perhaps, in a distinctive interplay of culture and technology. By culture, I mean less than the anthropological view, which includes

32

all "non-material" factors within the framework of a society, but more than the genteel view, which defines culture by some reference to refinement (e.g., the fine arts). By culture, I mean the efforts, of symbol makers to define, in a self-conscious way, the *meanings* of existence, and to find some justifications, moral and aesthetic, for those meanings. In this sense, culture guards the continuity of human experience. By technology—in a definition I will expand later— I mean the effort to transform nature for utilitarian purposes. In this sense, technology is always disruptive of traditional social forms and creates a crisis for culture. The ground on which the battle is fought is nature. In this paper, I want to deal largely with the vicissitudes of nature as it is reshaped by technology, and the vicissitudes of technology in its relation to society. To that extent, I have to forgo any extensive discussion of culture, though I shall return to that theme at the end.

III

What is nature? Any attempt at specific definition brings one up short against the protean quality of the term. Nature is used to denote the physical environment or the laws of matter, the "nature" of man (e.g., his "essence") and the "natural order" of descent (e.g., the family, in botany and society). We talk of "natural selection" as the fortuitous variations in individuals or species which assure survival, and "natural law" as the rules of right reason beyond institutional law.[6] In a satirical passage in *Rasselas*, Samuel Johnson has his young Prince of Abissinia meet a sage who, when asked to disclose the secret of happiness, tells him "to live according to his nature." Rasselas asks the philosopher how one sets about living according to nature and is told a string of generalities that expose the wise man's emptiness.[7]

For my purposes, I restrict the meaning of nature to two usages. The first is what in German—whose fine structure of prefixes allows one to multiply distinctions—is called the *Umwelt*, the organic and inorganic realms of the earth which are changed by man. This is the geography of the world, the environment. The second is what the Greeks called *physis*, or the order of things, which is discerned by man; this natural order is contrasted with *themis*, the moral order, and *nomos*, the legal order. For my purposes, then, nature is a realm outside of man whose designs are reworked by men.[8] In transforming nature, men seek to bring the timeless into time, to bring nature into history. The history of nature, then, is on two levels: the sequential transformations of the *Umwelt* as men seek to bend nature to their purposes, and the successive interpretations of *physis* as men seek to unravel the order of things.[9]

We begin with the *Umwelt*, and with myth. Man remakes nature for the simple and startling reason that man, of all living creatures, "natural man," is not at home in nature. Nature is not fitted to his needs. This is the insight first enunciated by Hesiod in *Works and Days*, and retold by Protagoras in the Platonic dialogues to spell out a moral about human society. The story, of course, is that of Prometheus and Epimetheus. The two brothers, foresight and hindsight, are charged by the gods with equipping the newly fashioned mortal creatures with "powers suitable to each kind." But, unaccountably—perhaps because of the pride of the younger to excel—Epimetheus asks the older for permission to do the job, and is given the task. He begins with the animals. Some are given strength and others speed, some receive weapons and others camouflage, some are given flight and others means of dwelling underground; those who live by devouring other animals are made less

prolific while their victims are endowed with fertility—
"the whole distribution on a principle of compensation,
being careful by these devices that no species should be destroyed."

But without forethought, Epimetheus squandered all his
available powers on the brute beasts, and none were left for
the human race. Prometheus, come to inspect the work,
"found the other animals well off for everything, but man
naked, unshod, unbedded, and unarmed, and already the
appointed day had come when man, too, was to emerge
from within the earth into the daylight." Prometheus therefore stole from Athena and Hephaestus the gift of skill in
the arts, together with fire. "In this way man acquired
sufficient resources to keep himself alive. . . ."[10] Nature
became refitted for man.

As Prometheus says, in the play of Aeschylus:

> I gave to mortals gift.
> I hunted out the secret source of fire.
> I filled a reed therewith,
> fire, the teacher of all arts to men,
> the great way through. . . .
>
> I, too, first brought beneath the yoke
> great beasts to serve the plow,
> to toil in mortals' stead. . . .
>
> Listen, and you shall find more cause for wonder.
> Best of all gifts I gave them was the gift of healing.
> For if one fell into a malady
> there was no drug to cure, no draught, or soothing
>   ointment. . . .
>
> The ways of divination I marked out for them,
> and they are many; how to know
> the waking vision from the idle dream;
> to read the sounds hard to discern;

35

the signs met on the road; . . .
So did I lead them on to knowledge
of the dark and riddling art.[11]

Natural goods are those we share with the animals, but cultivated or fabricated goods require the reworking of nature: the husbandry of soil and animals, the burning of the forests, the redirection of the rivers, the leveling of mountains. These demand acquired powers. The introduction of *techne* gives man a second nature, or different character, by extending his powers through adaptive skills and re-directive thought; it allows him to prefigure or imagine change and then seek to change the reality in accordance with the thought. The fruits of *techne* create a second world, a technical order which is superimposed on the natural order.

In the imagination of the Greeks, these stolen skills were powers of the gods, and with these powers man could begin that rope dance above the abyss which would transform him from "the kinship with the worm," in the phrase of Faust, to the godlike knowledge that partakes of the divine. Prometheus was punished, and, in the romantic imagination of Marx as well as Shelley, Prometheus was the eternal rebel who had dared to act for men. The paradox is that today the romantic imagination, having turned against *techne*, remains puzzled as to what to say about its primal hero. Most likely, the new shamans would say that the punishment was justified. But that is another story, for another day.

I jump now almost two thousand years, from Protagoras to the seventeenth century c.e., to a radically new way of looking at nature and of organizing thought, the rule of abstraction and number.[12]

Mythology, the first mode of depicting the world, is based on personification or metaphor. Nature is a creative or vital force ruling the *Umwelt*. In *Prometheus Bound* the characters are called Ocean, Force, and Violence, or in the later personification of the tides of destiny (we cannot escape metaphor in our speech) we find *Moira*, or Fate, and *Tyche*, or Chance, as the two principles which rule our lives. Through myth, metaphor, and characterization, we can dramatize our plights, and search for meaning in expressive symbolism; that is the virtue of the poetic mode. But with abstraction and number, we can state causal or functional relationships and predict the future states of, or manipulate, the world. Nature as *physis* is an order of things. The heart of the modern discovery is the word *method*. Nature is to be approached through a new method.

In terms of method, the first achievement, that of Galileo, was the simplification of nature. Galileo divided nature into the world of qualities and the world of quantities, the sensory order and the abstract order. All sensory qualities —color, sound, smell, and the like—were classified as secondary and relegated to subjective experience. In the physical world were the primary quantities of size, figure, number, position, motion, and mass, those properties which were capable of extension and mathematical interpretation. The worlds of poetry and physics, the idea of natural philosophy, were thus sundered.

Equally important was the contrast with the classical Aristotelian view which Thomas Aquinas had enlarged upon in medieval thought. Then the object of science was to discover the different purposes of things, their essence, their "whatness," and their qualitative distinction. But little attention was paid to the exactly measured *relations* between events or the *how* of things. In this first break with the past, measurement and relation became the mode. To do so Galileo shifted the focus of attention from specific objects to their abstract properties. One did not measure the fall of an object but mass, velocity, force, as the properties of bodies, and the relations among these properties. The elements of analytical abstraction replaced concrete things as the units of study.

The search for method, which was taken up by Descartes, was not just an effort at exactitude and measurement, but had a double purpose. The first was to raise the general intellectual powers of all men. An artistic ignoramus with a compass, said Descartes, can draw a more perfect circle than the greatest artist working free hand. The correct method would be to the mind what the compass is to the hand. And second, with this "compass of the mind" one could create a general method that would be a flawless instrument for the unlimited progress of the human mind in theoretical and practical knowledge of all kinds.[13]

That "dream of reason" is symbolized by a famous episode in Descartes's life. One day, confined to his room by a cold, he resolved to discard all beliefs that could not pass the test of reason. That night he had an intense vision, and with feverish speed he perfected the union of algebra and geometry—the complete correspondence between a realm of abstraction and a realm of real world space—that we now call analytical geometry.

As I considered the matter carefully, it gradually came to light that all those matters only are referred to mathematics in which order and measurement are investigated, and that it makes no difference whether it be in numbers, figures, stars, sounds or any other object that the question of measurement arises. I saw consequently that there must be some general science to explain that element as a whole which gives rise to problems about order and measurement, restricted as these are to no special subject matter. This, I perceived, was called universal mathematics. . . . To speak freely, I am convinced that it is a more powerful instrument of knowledge than any other that has been bequeathed to us by human agency, as being the source of all others.

Intoxicated by his vision and his success, Descartes declared, "Give me extension and motion, and I will construct the universe." And with Newton's mathematical method of computing the rates of motion, the calculus, the universe was constructed in exact, mathematical deductive terms.

The universal order [said Newton], symbolized henceforth by the law of gravitation, takes on a clear and positive meaning. This order is accessible to the mind, it is not preëstablished mysteriously, it is the most evident of all facts. From this it follows that the sole reality that can be accessible to our means of knowledge, matter, nature, appears to us as a tissue of properties, precisely ordered, and of which the connection can be expressed in terms of mathematics.

But what of design, purpose, value, *telos?* None. What we have with this "watershed," as Arthur Koestler has called it, is the desacralization, or—depending on one's temperament and values—the demystification, of nature.

The use of mathematics to discover the underlying order of the universe is, of course, not new. The Pythagoreans had sought to discern form, proportion, and pattern, expressed as relations, in the order of number—shapes and intervals which could be expressed in musical terms. The discovery that the pitch of a note depends on the length of the string which produces it and that concordant intervals in the scale are produced by simple numerical ratios, was a reduction of "quality to quantity," or the mathematicization of human experience. But with the Pythagoreans, as it was for Kepler, the mystical and scientific modes of experience were joined, each to illumine the other. With Galileo comes the radical separation. "He was utterly devoid of any mystical, contemplative learnings, in which the bitter passions could from time to time be resolved," Koestler writes; "he was unable to transcend himself and find refuge, as Kepler did in his darkest hours, in the cosmic mystery. He did not stand astride the watershed; Galileo is wholly and frighteningly modern."

With Galileo, physics becomes detached not only from mysticism but from "natural philosophy" as well. Aristotelian physics is valuative, reflecting a world conceived in hierarchic terms. The "highest" forms of motion are circular and rectilinear; and they occur only in the "heavenly" movements of the stars. The "earthly," sublunar world is endowed with motion of an inferior type. But, in the Newtonian world view, the idea of the heavens is detached from any ascending hierarchy of purposes as envisaged by Thomas Aquinas and is but a uniform, mathematical system on the single plane of motion.

And finally, in this mechanistic world view the world becomes sundered from the anthropomorphic image of a wise and loving Father, or the theological image of the being whose powers are so miraculous that he can create a world out of nothing—the doctrine of *creato ex nihilo*. As Spinoza put it in his geometry: "Nothing comes to pass in nature in contravention to her universal laws . . . she keeps a fixed and immutable order." And purpose and final cause? "There is no need to show at length that nature has no particular goal in view, and that final causes are mere human figments. . . . Whatsoever comes to pass, comes to pass by the will and eternal decree of God; what is, whatever comes to pass comes to pass according to rules which involve eternal necessity and truth."

Thus the order of nature, *physis*, is some vast *perpetuum mobile*, whose every point in time, including its future state, can be deduced mathematically from the fundamental principles of its mechanical action. Nature is a machine.

A third vision: man the inventor, the experimenter, the active, purposeful intervenor in the processes of nature to subordinate and bend the *Umwelt* to men's wills. This is a view which centers not on myth or mechanism but on man as an active reshaper of the world. The key word here is *activity*. If there is a radical difference between classical and modern views, it is the abandonment of the contemplative attitude toward nature, aesthetics, and thought, and the adoption of an activist attitude toward experience and the environment. All of modern epistemology is dependent on an activity theory of knowledge.

Our science and technology, Lynn White has written, grew out of the Christian view of nature as the dominion of men over the earth and other creatures. This may be too facile. Christian thought, as Clarence J. Glacken has observed, is also characterized by a *contemptus mundi*, a rejection of the earth as the dwelling place of man and an indifference to nature. And while Christian thought, to follow White, may have felt alien to an idea of a "sacred grove," and could be prodigal in its waste of nature, it did not have the impulse to rework God's designs for man's own purposes.[14]

It is in the period roughly from the end of the fifteenth century to the end of the seventeenth that the sources of the idea of man as a controller of nature and an active agent—in mind and in matter—began to take shape. For our purposes, there are two sources. One is a new and growing emphasis on *practical activity* and the emergence, during the Renaissance, of a group of remarkable men, artist-engineers, who united rational training with manual work to lay down the foundations of experimental science. These artist-engineers, as Edgar Zilsel remarks, not only painted pictures and built cathedrals but also constructed lifting engines, earthworks, canals, sluices, guns, and fortresses, discovered new pigments, formulated the geometrical laws of perspective, and invented new measuring tools for engineering and gunnery. Among these were Filippo Brunelleschi, the principal architect of the cathedral of Florence; the bronze founder Lorenzo Ghiberti; the architect Leon Battista Alberti; the architect and military engineer Francesco di Giorgio Martini; the incomparable Leonardo da Vinci, who drew maps, built canals, created weapons, and designed craft to submerge under the sea and others to fly through the air; the goldsmith, sculptor, gun constructor, and adventurer Benvenuto Cellini; and in Germany, Albrecht Dürer as a surveyor and cartographer. Related to them were instrument makers who supplied navigators, geodesists, and astronomers with their aids, and clock makers, cartographers, and military technicians who created new tools. Their knowledge was empirical, but they sought to systematize and generalize their experiences and recorded them in diaries or treatises for their colleagues and apprentices. From these experiences they sought to discover theories for general application. "Practice in painting must always be founded on sound theory," Leonardo wrote; and Alberti's book *On Painting*, based on geometry and optics, attempted to create a new "compass of vision" which would guide all use of perspective in art, as almost two hundred years later Descartes would seek a method for the compass of the mind.[15]

The second source of the "activity principle" derives from the Cartesian revolution in concepts. The contemplative tradition of mind, going back to Greek and Christian thought, saw the human being as an observer passively regarding a world unfolding in front of him; knowledge was a copy, so to speak, of a picture of what that outside world was like. So long as science followed Aristotelian realism, and the world was just what its picture seemed to the mind, there were no difficulties. But Galileo offered a different conception. Water, for example, was not just a wet, formless fluid of varying temperature that one could observe, but a number of particles of matter whose motion followed definite laws. And for Descartes, the qualities of wetness and coolness were not just properties of water, but generalized concepts in the mind that perceived the water. Knowledge of nature, or of the world, thus depended not on immediate experience but on the axioms of geometry and the categories of mind.

The world of appearance and the substructures of reality had come apart. Yet the two could be joined, since what one knows, as Kant developed the idea, is a function of the selective categories by which mind relates different worlds of fact and appearance. Mind organizes perception through selective scanning, takes different attributes and properties of objects or events, groups these together for the purposes of analysis and comparison, and organizes them into conceptual systems whose use is tested by empirical application. Mind is thus an active agent in the making of judgments about reality. But these judgments, to be valid, have to be confirmed through the prediction of consequences. To know the world, one has to test it. In the older meanings of the words, *theria* meant to see, and practice, to do; but necessarily, activity joins the two. In this way, rationalism and empiricism are interlocked through *praxis*.

Thus, a new world view: the emphasis on practical activity and on the role of mind as an active formulator of plans reworking the categories of nature. As Descartes observed in Part V of the *Discourse on Method,* what he was seeking for was a science and technique which would make men "the masters and possessors of Nature."

The spread of a world view requires a prophet. The prophet provides the passion for a view; he makes claims (usually extravagant) and gains attention; he codifies the doctrine and simplifies the argument—in short, he provides a formula for quick understanding and a moral rationale for necessary justification. Of this new world view, there were two significant prophets: Francis Bacon and Karl Marx.

Francis Bacon, lawyer, politician, essayist—at one time Lord Chancellor of England in 1618—was not a scientist. As Alexandre Koyré has remarked: if all of Bacon's writings were to be removed from history, not a single scientific concept, not a single scientific result, would be lost. Yet science might have been different without him, or without someone of his literary gifts. For what Bacon did was to formulate the modern credo: of science as the endless pursuit of knowledge; of the experiment as the distinctive mode of science; and of utility as the goal and purpose of science.

Of the endless pursuit of knowledge Edgar Zilsel has observed:

The modern scientist looks upon science as a great building erected stone by stone through the work of his predecessors and his contemporary fellow-scientists, a structure that will be continued but never be completed by his successors. . . . [Science] is regarded as the product of a cooperation for non-personal ends, a cooperation in which all scientists of the past, the present and the future have a part. Today this idea or ideal seems almost self-evident. Yet no Brahmanic, Buddhistic, Moslem or Catholic Scholastic, no Confucian scholar or Renaissance Humanist, no philosopher or rhetor of classical antiquity ever achieved it. It is a special characteristic of the scientific spirit and of modern western civilization. It appeared for the first time fully developed in the works of Francis Bacon.[16]

Against scholasticism, Bacon emphasized the importance of the experiment. He was himself an enthusiastic experimenter, so much so that he died from a cold he caught while stuffing a dead chicken with snow during an experiment on the preservation of food.[17] Bacon believed in the experiment as the foundation of inductive thinking. Science today has a less exalted view of induction, and the most sophisticated philosophies of science mock its possibilities at all; yet the idea of the experiment remains as the necessary condition for the disconfirmation of hypotheses. As Charles C. Gillespie noted: ". . . Bacon's emphasis on experiment did shape the style of science. So strongly did it do so that the term 'experimental science,' has become practically a synonym for 'modern science,' and nothing so clearly differentiates post-seventeenth century science from that of the Renaissance, or of Greece, as the role of experiment."[18]

But experiment, for Bacon, was linked to a specific purpose: "The true and lawful goal of the sciences is none other than this: that human life be endowed with new discoveries and powers." As with Leonardo, one finds in Bacon an admiration for the inventor and the experimenter, and a contempt for the pretensions of authority (at least that of the past; at the courts of Elizabeth and James he was a toady and was rewarded for his obsequiousness with a peerage: one can say that he rendered unto science that which was science's and to Caesar that which Caesar demanded; when personal survival is at stake, it is difficult to be harsh on the courtier). It was to the mechanical arts, to industry and seafaring, to the practical men, that Bacon went for the source of knowledge. The scholastics and humanists, he declared, have merely repeated the sayings of the past. Only in the mechanical arts has knowledge been furthered since antiquity. In Bacon's philosophy, there is no choice to be made between basic and applied science. On the contrary, applied science is by definition basic; it is the object of the search.[19]

All of this is summed up in his last book, a utopian fable, *The New Atlantis.* The choice of form and the selection of place are certainly not fortuitous. The form is a voyage of discovery. Fresh to the times are the great voyages which have expanded man's geographic vision, and for Bacon these voyages symbolize the broadening of intellectual horizons as well. The place is Atlantis, the lost continent of the cosmological allegory of the *Timaeus*, the place where Plato introduced the notion of a demiurge as artisan-deity, the one who creates an orderly universe out of recalcitrant materials.

In Bacon's utopia, the philosopher is no longer king; his place has been taken by the research scientist. The most important building in this frangible land of Bensalem is Salomon's House, not a church but a research institute, "the noblest foundation . . . that ever was upon the earth and the lantern of this kingdom." Salomon's House, or the College of the Six Days Works, was a state institution "instituted for the production of great and marvelous works for the benefit of man." Of the thirty-five pages which make up the complete work, ten are devoted to the listing of the marvels of inventions that have been gathered in this treasure house. The Master of the House, speaking to his visitors, declares: We imitate the flight of birds: for we have some degree of flying in the air; we have ships and boats for going under water. There are engines for insulation, refrigeration and conservation. Animals are made bigger or smaller than their natural size. There is a perspective house to demonstrate light, a sound house for music, and engine house to imitate motions, a mathematical house, and "houses of deceits of the senses, where we represent all manner of . . . false apparitions, impostures, illusions and their failures."

For Seneca (who faced similar political problems) and the Stoics, the exploration of nature was a means of escaping the miseries of life. For Bacon, the purpose of knowledge was utility, to increase happiness and to mitigate suffering. In summing up the marvels, the master of Salomon's House bespeaks the boundless ambition of all technological utopias. "The end of our foundation," he says, "is the knowledge of causes and the secret motions of things . . . the enlarging of the bounds of human empire to the effecting of all things possible."[20]

For Marx, like Protagoras, man, too, is the measure of all things. What is striking is his fierce contempt for the romantic cult of nature and his denigration of the sentimental idylls—"drivel," he called them—about the land and the forests. In *The German Ideology* Marx mocks the utopian socialists who see a harmonious unity in nature. Where the "true socialist" sees "gay flowers . . . tall and stately oaks . . . their satisfactions lie in their life, their growth, their blossoming," Marx remarks, " 'Man' could [also] observe . . . the bitterest competition among plants and animals; he could see . . . in his 'forest of tall and stately oaks,' how these tall and stately capitalists consume the nutriment of the tiny shrubs. . . ."[21]

This sarcasm is all the more remarkable since Marx came to his own philosophy through the forest of Feuerbach, his mentor, who preached a sensual cult of nature and proclaimed Man rather than God as the center of the world. In an essay written in 1950, a review of a book by Daumer entitled *The Religion of a New Age,* Marx renounced any kind of Feuerbachian cult of "Man" or "Nature." Where the author tries to establish a "natural religion" in modern form, based on reverence for nature, Marx counterposes science:

[In the work] there is no question, of course, of modern sciences, which, with modern industry, have revolutionized the whole of nature and put an end to man's childish attitude towards nature as well as to other forms of childishness. . .[22]

For Marx, nature is blind, nature is necessity. Against blind nature is conscious man, who, in his growing consciousness, is now able to plan and direct his future. Against necessity, the need to toil, is History, a new demiurge to wrest plenty from a recalcitrant nature. In the movement of History comes the point, in the dithyrambic phrase of Engels, where there is the "leap from the realm of necessity into the realm of freedom," the "genuinely human" period to which all recorded history has been an antechamber. As Marx writes in *Capital:*

Just as the savage must wrestle with nature to satisfy his wants, to maintain and reproduce life, so must civilized man, and he must do so in all social formations and under all possible modes of production. With his development this realm of physical necessity expands as a result of his wants; but at the same time, the forces of production which satisfy these wants also increase. Freedom in this field can only consist in socialized men, the associated producers, rationally regulating their interchange with nature, bringing it under their common control, instead of being ruled by it as the blind forces of nature. . . But it nonetheless still remains a realm of necessity. Beyond it begins that development of human energy which is an end in itself, the true realm of freedom, which, however, can blossom forth only with this realm of necessity as its basis.[23]

The fulcrum of all this is human needs. Cartesian rationalism and Kantian idealism had created a theory of mind as activity, against the old, passive materialism. But, for Marx, consciousness alone, though it can interpret, that is, fully prefigure, a world—this is what he meant by the "realization" of philosophy—cannot change the world. Man does so because he has needs which can be satisfied only by transforming nature. With Marx, the activity theory of mind is reworked as an activity theory of material needs and powers.

For Marx, the agency of man's power is technology. History is the history of the forces of production, and the significance of the modern era is the extraordinary power which a new class in history, the bourgeoisie, has achieved through *techne*. In the *Communist Manifesto*, the herald of the socialist revolution, Marx writes a startling panegyric to capitalism:

The bourgeoisie was the first to show us what human activity is capable of achieving. It has executed works more marvellous than the building of Egyptian pyramids, Roman aqueducts, and Gothic cathedrals. . . .
During its reign of scarce a century, the bourgeoisie has created more powerful, more stupendous forces of production than all preceding generations rolled into one. The subjugation of the forces of nature, the invention of machinery, the application of chemistry to industry and agriculture, steamships, railways, electric telegraphs, the clearing of whole continents for cultivation, the making of navigable waterways, huge populations springing up as if by magic out of the earth—what earlier generations had the remotest inkling that such productive powers slumbered within the womb of social labour?[24]

"The bourgeoisie cannot exist," wrote Marx, "without

49

incessantly revolutionizing the instruments of production. . . . That which characterizes the bourgeois epoch in contradistinction to all others is a continuous transformation of production." These forces of production, however, are held in check by the property owners, those who can only be superfluous in the new society, and this contradiction sets limits on the development of the economy and the productive forces themselves. Remove the bourgeoisie, and all the brakes on production will be released.

What is clear from all this is Marx's faith in technology as the royal road to utopia; technology had solved the single problem which had kept the majority of men in bondage throughout human history and which was responsible for almost all the ills of the world: economic scarcity. Because of scarcity, each man is pitted against the other, and man becomes wolf to man. Scarcity is the condition of the state of nature, which is why, says Marx, Hobbes was right that the state of nature is a war of one against all.

The "end of history" is the substitution of a conscious social order for a natural order. The unfettered reign of technology is the foundation of abundance, the condition for the reduction, if not the end, of necessity.[25]

IV

Technology, like art, is a soaring exercise of the human imagination. Art is the aesthetic ordering of experience to express meanings in symbolic terms, and the reordering of nature—the qualities of space and time—in new perceptual and material form. Art is an end in itself; its values are intrinsic. Technology is the instrumental ordering of human experience within a logic of efficient means, and the direction of nature to use its powers for material gain. But art and technology are not separate realms walled off from the other. Art employs *techne*, but for its own ends. *Techne*, too, is a form of art that bridges culture and social structure, and in the process reshapes both.[26]

This is to see technology in its essence. But one may understand it better, perhaps, by looking at the dimensions of its existence. For my purposes, I will specify five dimensions in order to see how technology transforms both culture and social structure.

1. *Function.* Technology begins with an aesthetic idea: that the shape and structure of an object—a building, a vehicle, a machine—are dictated by its function. Nature is a guide, only to the extent that it is efficient. Design and form are no longer ends in themselves. Tradition is no justification for the repetition of designs. Form is not the unfolding of an immanent aesthetic logic—such as the musical forms of the eighteenth to early-twentieth centuries. There is no dialogue with the past. It is no accident that the adherents of a machine aesthetic in the early-twentieth century, in Italy and Russia, flaunted the name Futurism.

2. *Energy.* Technology is the replacement of natural sources of power by created power beyond all past artistic imagination. Leonardo made designs for submarines and air-conditioning machines, but he could not imagine any sources of power other than what his eyes could behold: human muscles or animal strength, the power generated by wind and falling water. Visionaries of the seventeenth century talked grandiosely of mechanized agriculture, but their giant combines were to be driven by windmills and thus could not work. Energy drives objects—ships, cars, planes, lathes, machines—to speeds thousands of times faster than the winds, which were the limits of the "natural" imagination; creates light, heat, and cold, extending the places where people can live and the time of the diurnal

cycle; lifts weights to great heights, permitting the erection of scrapers of the skies and multiplying the densities of an area. The skyscraper lighted at night is as much the technological symbol of the modern city as the cathedral was the emblem of medieval religious life.

3. *Fabrication.* In its oldest terms, technology is the craft or scientific knowledge which specifies ways of doing things in a reproducible manner.[27] The replication of items from templates or dies is an ancient art; its most frequent example is coinage. But the essence of technology is that, owing to the standardization of skills and the standarization of objects, its reproduction is much cheaper than its invention or development. Modern technological fabrication introduces two different factors: the replacement of manual labor and artisan skills by programmed machines; and the incredible rapidity of reproduction—the printing of a million newspaper copies per night—which is the difference in scale.

4. *Communication and Control.* Just as no one before the eighteenth century could imagine the new kinds of energy to come, so, well into the nineteenth century, could no one imagine—even with the coding of messages into dots and dashes, as in telegraphy—the locking of binary digits with electricity or the amplification of ethereal waves, which has produced modern communication and control systems. With telephone, radio, television, and satellite communication, one person can talk to another in any part of the globe, or one person can be seen by hundreds of millions of persons at the same moment. With programmed instructions—through the maze of circuits at nanoseconds speed—we have control mechanisms that switch trains, guide planes, run automated machinery, compute figures, process data, simulate the movement of the stars, and correct for error both human and machine. And an odd phrase

sums it all up: these are all done in "real time."

5. *Algorithms.* Technology is clearly more than the physical manipulation of nature. There is an "intellectual technology" as well. An algorithm is a "decision rule," a judgment of one or another alternative course to be taken, under varying conditions, to solve a problem. In this sense we have technology whenever we can substitute algorithms for human judgment.

We have here a continuum with classical technology, but it has been transposed to a new qualitative level. Physical technology—the machine—replaced human power at the *manual* level of raw muscle power or finger dexterity or repetition of tasks; the new intellectual technology—as embodied in a computer program or a numerically controlled machine tool—substitutes algorithms for human *judgments,* where these can be formalized. To this extent, the new intellectual technology marks the last half of the twentieth century, as the machine was the symbol of the first half.

Beyond this is a larger dream, the formalization of a theory of choice through stochastic, probabilistic, and deterministic methods—the applied mathematics of Markov chains, Monte Carlo randomization, or the minimax propositions of game theory. If the computer is the "tool," then "decision theory" is its master. Just as Pascal sought to throw dice with God, or the physiocrats to draw a perfect grid to array all economic exchanges among men, so the decision theorists, and the new intellectual technology, seek their own *tableau entier*—the compass of rationality itself.

We say knowingly that technology has transformed the world, and we feel, apprehensively, that we are on a steeply rising "exponential curve" of change, so that the historical transition to new levels of technological power all over the world creates a crisis of transformation. But it is not always clear what has been transformed. To say vaguely "our lives," or flashily that we are experiencing "future shock," is of little help in understanding the character of that transformation. Even the fashionable phrase "the acceleration in the pace of change" is of little help, for it rarely specifies change in *what*. If one thinks of inventions that change the character of daily life, it is not at all clear that our generation is experiencing "more" change (and how does one measure it?) than previous ones. As Mervyn Jones points out, "A man born in 1800 and dying in 1860 would have seen the coming of the railway, the steamship, the telegraph, gas lighting, factory-made clothing and furniture. A man born in 1860 and dying in 1920 would have seen the telephone, electric light, the car and the lorry, the aeroplane, radio and the cinema."[28] Are television and the computer any more "shocking" in their impact on our lives than the inventions which changed the lives of our immediate forebears?

Yet technology has transformed our lives in more radical, yet more subtle, forms than the marvelous "gadgets" that we can see. From a sociological point of view, the effects are twofold: a change in the "axis" of economics, and an increase in the interaction—and consequently the "moral density," to use Durkheim's phrase—among men. These two constitute a definition of "modernity."

For most of the thousands of years of human existence, the level of production dictated the levels of consumption; in the shorthand of economics, supply shaped demand. The returns from nature were small, and, from any single unit of production—in almost all cases, land—they were also diminishing, so that after a period of time men would either have to wait and let the forests and land lie fallow, to be replenished slowly by nature, or move on in search of new lands. Given the scarce returns, men fought all the more bitterly for their share, and the major means of amassing wealth were war and plunder. With technology, this situation was changed. One finds an increasing proportion of returns for a given effort—we call this productivity—and, more subtly, a shift in the axis of economics from supply to demand. What men now want begins to affect the levels of production and to dictate the different kinds of items to be produced. It is this change from supply to demand which, in the intellectual sphere, creates modern economics. In the social world, it creates an entirely new attitude toward the world, wealth, and happiness; the three become defined by the sharp rise in the standard of living of the masses of men in the world.[29]

The second change is the breakdown of segmentation and the enlargement of the boundaries of society—the creation, in sociological terms, of the mass society. The mass society is not just a large society. Czarist Russia and Imperial China were large-land-mass societies, but they were segmented; each village was much like any other, repeating within its bounds the same kind of social structure. Segmentation disappears not with the growth of population but with the increase in the degree of social intercourse or contact—or, more technically, the rate of interaction, per unit of time, between men. Physical density is the number of persons per unit of space. Sociological density is the number of connections among people at a given time. When the Constitution was first adopted, the population of the United States was four million, and New York City, then the capital, held a total of 30,000 persons. Today there are more than 200 million persons in the country, and the metropolitan areas within which people live and work commonly hold from 5 to 20 million persons. Yet the change in the country is not just the increase in numbers, but the quantum jump in interactions as well. As against the life of individuals at the turn of the nineteenth century, if we think of how many different individuals we meet (simply to speak to), how many individuals we know (those we encounter socially), and how many we know of (in order to recognize their names and be responsive to some comment about them), we get some sense of the change in the scale of our lives.

This axial change in the character of social relations brings with it two contrary changes in the areas of work and ideas: differentiation in the social structure and syncretism in the culture.

In the social structure—the realm of the economy, occupations, and stratification system of the society—the increase in interaction (between persons, firms, cities, regions) leads inevitably to competition. In earlier times, with few resources and growth limited, such a zero-sum situation—in which one could gain only at the expense of the other—led to plunder and exploitation. But, in an economic world where everybody can gain (albeit in differential amounts), competition leads to specialization, differentiation, and interdependence. "We can say," Durkheim remarked, "that the progress of the division of labor is in direct ratio to the moral or dynamic density of the society." The condition of efficiency, which is the basis of economic progress, is the growth of specialization and the narrower focusing of tasks and skills. In the area of work, intellectual and otherwise, man becomes a smaller and smaller part of a larger and larger whole. He becomes defined by his role.

In culture, the situation is remarkably reversed. In the traditional society the ideas one has, the beliefs one accepts, the arts one beholds are all within a bounded space. Modernity bursts the walls. Everything is now available. Hindu mantras and Tantric mandalas, Japanese prints and African sculptures, Eskimo music and Indian ragas all jostle with one another in "real time" within the confines of Western homes. Not since the age of Constantine has the world seen so many strange gods mingling in the meditative consciousness of the middle-class mind.

In principle, much of this is not new. What is distinctive is the change of scale. If any single principle dominates our life, it is that. All that we once knew played out on the scale of the Greek polis is now played out in the dimensions of the entire world. Scale creates two effects: One, it extends the range of control from a center of power. (What is Stalin, an unknown wit remarked, if not Genghis Khan with a telephone?) And two, when linear extension reaches certain thresholds, unsettling changes ensue. A university of fifty thousand students may bear the same name that university had thirty years before with five thousand students, but it is no longer the same university. A change in quantities is a change in quality; a change in scale is a change in institutional form.

This principle was laid down more than three hundred and fifty years ago by Galileo, in the square-cube law. As something doubles in size, its volume will triple; but then its shape will also be different. As the biologist D'Arcy Wentworth Thompson pointed out:

[Galileo] said that if we tried building ships, palaces or temples of enormous size, beams and bolts would cease to hold together; nor can Nature grow a tree nor construct an animal beyond a certain size while retaining the proportions and employing the materials which suffice in the case of a smaller structure. The thing will fall to pieces of its own weight unless we either change its relative proportions . . . or else we must find a new material, harder and stronger than was used before.[30]

The major question which confronts the twenty-first century is the question of the limits of scale. The technological revolution, as I have indicated, consisted in the availability of huge amounts of energy at cheaper cost, more control of the circumstances of production, and faster communication. Each development increased the effectiveness of the other two. All three factors increased the speed of performing large-scale operations. Yet, as John Von Neumann pointed out seventeen years ago:

. . . throughout the development, increased speed did not so much shorten time requirements of processes as extend the areas of the earth affected by them. The reason is clear. Since most *time* scales are fixed by human reaction times, habits, and other physiological and psychological processes, the effect of the increased speed of technological processes was to enlarge the *size* of units—political, organizational, economic, and cultural—affected by technological operations. That is, instead of performing the same operations as before in less time, now larger-scale operations were performed in the same time. This important evolution has a natural limit, that of the earth's actual size. The limit is now being reached, or at least closely approached.[31]

We first reached that limit in the geopolitical military sphere. In previous epochs, geography could provide an escape. Both Napoleon and Hitler became bogged down in the large land mass of Russia, though by 1940 even the larger countries of continental Western Europe were inadequate as military units. Since 1945, and particularly with the development of intercontinental ballistic missiles with multiple warheads, space and distance offer no effective cover or retreat in any part of the earth. As Von Neumann presciently observed in 1955, "The effectiveness of offensive weapons is such as to stultify all plausible defensive time scales." To recall an old phrase of Winston Churchill's, the equilibrium of power has become the balance of terror.

We are told that we will soon be reaching the limits of resources on a world scale. As a theorem, this is a tautology; as a practical fact, the time scale is elusive. The studies that have sounded the warning are faulty in their methodology.[32] They take little account of economics and the principles of relative prices and substitutability. The resources that may be available from untapped areas (e.g., the ocean bottoms, the large land areas of the Amazon, central Asia, Siberia, western China, the Antarctic) are uncharted. The degree of technological innovation that allows us to substitute light and cheaper materials for heavy and cumbersome ones (e.g., the role of semiconductors such as transistors in radios, television sets, and computers)—the entire range of miniaturization—is an unknown factor. And yet the warning is useful. More than two thirds of the world is in a pre-industrial phase wherein 60 per cent or more of the labor force is engaged in agriculture, timber, fishing, or mining, games against nature whose returns inevitably diminish. These countries obviously want to industrialize in order to raise their standards of living. The question we must confront is whether our resources are sufficient for such a task, whether new technologies can provide a more economical way, or whether the new industrialization and standard of living must come from some redistribution of the wealth of the advanced industrial countries of the world.

It is possible that we are reaching that limit of scale in technological terms. We have increased our speeds of communication by a factor of 10^7, our speeds of travel by 10^2, our speeds of computer operation by 10^6, and our energy resources by 10^3 in the past century.[33] But all exponential growth reaches an asymptote, the ceiling limit, where it levels off. In terrestrial speed, there is a natural limit of sixteen thousand miles an hour, since any higher speed throws a vehicle out of the earth's orbit. With aircraft, we are questioning whether we should go above supersonic speed because of the danger it might present to the earth's atmosphere or to human noise tolerance on the ground. In communication around the world, we have already approached, in telephonic, radio, and television communication, "real time," and the technological problems are primarily those of expanding the number of bands of communication to permit more people to enjoy that use.

In a fundamental sense, the space-time framework of the world *oikoumene* is now almost set. Transportation and communication bind the world as closely together today as the Greek polis of twenty-five hundred years ago. The major sociological problem created by that technology is what happens when all segmentation breaks down and a quantum jump in human interaction takes place. How will we manage when each and every part of the globe becomes accessible to every person? It was once suggested that 7×720 (5,040) citizens was the optimum size for the city-state. (If half a day a year is needed to maintain contact with a relatively good friend, there is a ceiling of 720 persons with whom we could have personal interaction.) Athens, the largest of the ancient Greek city-states, had, at the highest estimate, forty thousand male citizens, and a quorum in the assembly was fixed at six thousand. The number of adult citizens of New Zealand is around 30 times that in Athens, of the Netherlands 100 times, of France 500, of the United States about 2,500, and of India, the largest representative democracy in the world, about 5,000 or 6,000 times.[34] In the face of these numbers, what does participation mean? What is the character of human contact? What are the limits of human comprehension?

V

The rhetoric of apocalypse haunts our times. Given the recurrence of the Day of Wrath in the Western imagination—when the seven seals are opened and the seven vials pour forth—it may be that great acts of guilt provoke fears of retribution which are projected heavenward as mighty punishments of men. A little more than a decade ago we had the apocalyptic specter (whose reality content was indeed frightening) of a nuclear holocaust, and there was a flood of predictions that a nuclear war was a statistical certainty before the end of the decade. That apocalypse has receded, and other guilts produce other fears. Today it is the ecological crisis, and we find, like the drumroll of Revelation 14 to 16 recording the plagues: *The Doomsday Book, Terracide, Our Plundered Planet, The Chasm Ahead, The Hungry Planet*, and so on.[35]

In the demonology of the time, "the great whore" is technology. It has profaned Mother Nature, it has stripped away the mysteries, it has substituted for the natural environment an artificial environment in which man cannot feel at home.[36] The modern heresy, in the thinking of Jacques Ellul, the French social philosopher whose writing has been the strongest influence in shaping this school of thought, has been to enshrine *la technique* as the ruling principle of society.

Ellul defines technique as:

the translation into action of man's concern to master things by means of reason, to account for what is subconscious, make quantitative what is qualitative, make clear and precise the outlines of nature, take hold of chaos and put order into it.

Technique, by its power, takes over the government:

> Theoretically our politicians are at the center of the machinery, but actually they are being progressively eliminated by it. Our statesmen are important satellites of the machine, which, with all its parts and techniques, apparently functions as well without them.

Technique is a new morality which "has placed itself beyond good and evil and has such power and autonomy [that] it in turn has become the judge of what is moral, the creator of a new morality." We have here a new demiurge, an "unnatural" and "blind" logos that in the end enslaves man himself:

> When technique enters into the realm of social life, it collides ceaselessly with the human being. . . . Technique requires predictability and, no less, exactness of prediction. It is necessary, then, that technique prevail over the human being. For technique, this is a matter of life and death. Technique must reduce man to a technical animal, the king of the slaves of technique.[37]

Ellul has painted a reified world in which *la technique* is endowed with anthropomorphic and demonological attributes. (Milton's Satan, someone remarked, is Prometheus with Christian theology.) Many of the criticisms of technology today remind one of Goethe, who rejected Newton's optics on the ground that the microscope and telescope distorted the human scale and confused the mind. The point is well taken, if there is confusion of realms. What the eye can see unaided, and must respond to, is different from the microcosm below and the macrocosm beyond. Necessary distinctions have to be maintained. The difficulty today is that it is the critics of technology who absolutize the dilemmas and have no answers, short of the apocalyptic solutions that sound like the familiar comedy routine "Stop the world, I want to get off."

Against such cosmic anguish, one feels almost apologetic for mundane answers. But after the existentialist spasm, there remain the dull and unyielding problems of ordinary, daily life. The point is that technology, or technique, does not have a life of its own. There is no immanent logic of technology, no "imperative" that must be obeyed. Ellul has written: "Technique is a means with a set of rules for the game. . . . There is but one method for its use, one possibility."[38]

But this is patently not so if one distinguishes between technology and the social "support system" in which it is embedded.[39] The automobile and the highway network form a technological system; the way this system is used is a question of social organization. And the relation between the two can vary considerably. We can have a social system that emphasizes the private use of the automobile; money is then spent to provide parking and other facilities necessary to that purpose. On the other hand, arguing that an automobile is a capital expenditure whose "down time" is quite large, and that twenty feet of street space for a single person in one vehicle is a large social waste, we could penalize private auto use and have only a rental and taxi system that would substantially reduce the necessary number of cars. The same technology is compatible with a variety of social organizations, and we choose the one we want to use.

One should also distinguish between technology and the accounting system that allocates costs. Until recently, the social costs generated by different technologies have not been borne by the individuals or firms responsible for them, because the criterion of social accountability was not used. Today that is changing. The technology of the internal-combustion engine is being modified because the government now insists that the pollution it generates be reduced. And the technology is being changed. The energy crisis we face is less a physical shortage than the result of new demands—by consumers, and by socially minded individuals for a different kind of technological use of fuels. If we could burn the high-sulphur fuels used until a few years ago, there would be less of an energy crisis; but there would be more pollution. Here, too, the problem is one of costs and choice.

The source of our predicament is not the "imperatives" of technology but a lack of decision mechanisms for choosing the kinds of technology and social support patterns we want. The venerated teacher of philosophy at City College Morris Raphael Cohen used to pose a question to his students in moral philosophy: If a Moloch God were to offer the human race an invention that would enormously increase each individual's freedom and mobility, but demanded the human sacrifice of thirty-thousand lives (the going price at the time), would you take it? That invention, of course, was the automobile. But we had no mechanisms for assessing its effects and planning for the control of its use. Two hundred years ago, no one "voted" for our present industrial system, as men voted for a polity or a constitution. To this extent, the phrase "the industrial revolution" is deceptive, for there was no single moment when people could decide, as they did politically in 1789 or 1793 or 1917, for or against the new system. And yet today, with our increased awareness of alternates and consequences, we are beginning to make those choices. We can do this by technology assessment, and by social policy which either penalizes or encourages a technological development (e.g., the kind of energy we use) through the mechanism of taxes and subsidy.

A good deal of our intellectual difficulty stems from the way we conceive of society. Émile Durkheim, one of the founding fathers of modern sociology, contributed to this difficulty by saying that society exists *sui generis*, meaning that it could not be reduced to psychological factors. In a crucial sense he was right, but in his formulation he pictured society as an entity, a collective conscience outside the individual, acting as an external constraint on his behavior. And this lent itself to the romantic dualism of the individual versus society.

Society is *sui generis*, a level of complex organizations created by the degree of interdependence and the multiplicity of ties among men. A traffic jam, as Thomas Schelling has pointed out, is best analyzed not in terms of the individual pathologies of the drivers, but by considering the layout of roads, the pattern of flow into and out of the city, the congestion at particular times because of work scheduling, and so on. Society is not some external artifact, but *a set of social arrangements, created by men,* to regulate normatively the exchange of wants and satisfactions.

The order of society differs from the order of nature. Nature is "out there," without *telos*, and men must discern its binding and constraining laws to refit the world. Society is a moral order, defined by consciousness and purpose, and justified by its ability to satisfy men's needs, material and transcendental. Society is a design that, as men become more and more conscious of its consequences and effects, is subject to reordering and rearrangement in the effort to solve its quandaries. It is a social contract, made not in the past but in the present, in which the constructed rules are obeyed if they seem fair and just.

The problems of modern society arise from its increasing complexity and interdependence—the multiplication of interaction and the spread of syncretism—as old segmentations break down and new arrangements are needed. The resolution of the problem is twofold: to create political and administrative structures that are responsive to the new scales, and to develop a more comprehensive or coherent creed that diverse men can share. The prescription is easy. It is the exegesis, as the listener to Rabbi Hillel finally understood, that is difficult.[40]

VI

I return to my original question: Is the evident expansion of man's powers a measure of progress; and how do we talk to the Greeks and they to us? I began this discussion with the myth as told by Protagoras, but I did not finish it then; we now return to it.

Following the theft of fire, "man had a share in the portion of the Gods." But he soon found that *techne* does not create civilized life. When men gathered in communities, they injured one another for want of political skill. As Protagoras recounts it:

Zeus therefore, fearing the total destruction of our race, sent Hermes to impart to men the qualities of respect for others and a sense of justice, so as to bring order into our cities and create a bond of friendship and union.

Hermes asked Zeus in what manner he was to bestow these gifts on men. "Shall I distribute them as the arts were distributed—that is, on the principle that one trained doctor suffices for many laymen, and so with the other experts? Shall I distribute justice and respect for their fellows in this way, or to all alike?"

"To all," said Zeus. "Let all have their share. There could never be cities if only a few shared in these virtues, as in the arts. Moreover, you must lay it down as my law that if anyone is incapable of acquiring his share of these two virtues he shall be put to death as a plague to the city."[41]

What we have here, in Homer's earlier terms, is the contrast between *techne* and *themis*. *Techne* enables us to conquer nature; it is essential to the *economic* life. But *themis*, the marriage of conscience and honorable conduct, is the principle of civilized life. "*Themis*," as James Redfield puts it, "is the characteristic human good, and man is distinguished from the feral savage by his ability to live in a society." In Homer, *themis* is primary, and *techne* secondary.[42]

Hegel interpreted Plato's myths as representing a necessary stage in the education of the human race—the childhood phase—which conceptual knowledge could discard as soon as philosophy had grown up. But it should be clear by now that the image of childhood, as used by Hegel and Marx, is meaningless. We are not much ahead of the Greeks in the formulation of our problems or in our wisdom for solving them. In what sense, then, are we alike, and in what sense different?

Society, I would say, should be regarded as having three analytically distinct dimensions—the culture, the polity, and the social structure—each characterized by a different axial principle and each possessing a different historical rhythm.

Culture embraces the areas of expressive symbolism (painting, poetry, fiction) which seeks to explore these meanings in imaginative form; the codes of guidance for behavior, which spell out the limits, prescriptive and prohibitive, of moral conduct; and the character structure of individuals as they integrate these dimensions in their daily lives. But the themes of culture are the existential questions that face all human beings at all times in the consciousness of history—how one meets death, the nature of loyalty and obligation, the character of tragedy, the definition of heroism, the redemptiveness of love—and there is a principle of limited possibilities in the modes of response. The principle of culture, thus, is a *ricorso*, returning, not in its forms but in its concerns, to the same *essential* modalities that represent the finitude of human existence.

The polity, which is the regulation of conflict under the constitutive principle of justice, involves the different forms of authority by which men seek to rule themselves: oligarchy and democracy, elite and mass, centralization and decentralization, rule and consent. The polity is *mimesis*, in which the forms are known and men choose those appropriate to their times.

The social structure—the realm of the economy, technology, and occupational system—is *epigenetic*. It is linear, cumulative, and quantitative, for there are specific rules for the process of growth and differentiation.

To the extent that man becomes more and more independent of nature, he can choose and construct the kind of society he wants. Yet he is constrained by the axial facts that each societal realm has a different rhythm of change and that not all forms are compatible with each other.

If one asks, then, in what ways we have advanced beyond the Greeks, we know that our time-space perceptions of the earth have changed, for we have both speed and the view from the air, which the Greeks never knew. And in the power to transform nature and extend the range of man's political life, we live on a scale they would not have been able to understand. Our social structures, then, are vastly different, transformed as they are by technology. Our polities resemble each other in their predicaments (one has only to read Thucydides to be struck by the resemblances), but the problems today are greatly distended by the influence of numbers of persons, and the simultaneity of issues. Yet when we read the major chorus of Sophocles' *Antigone*—"Wonders are many," ending with the antistrophe "the craft of his engines has passed his dream/In haste to the good or evil goal"—we know that, with all the celebration of man's powers to navigate the seas and to domesticate the earth, man without justice and righteousness ("no city hath he who, for his rashness, dwells with sin") is his own enemy, and that we are all, over the span of the millennia, human, all too human.

VII

In the *ricorsi* of human existence, there have been recurrent cycles of optimism and despair. In the Greek world one finds Hesiod regarding society as corrupt, nature as recalcitrant, and history as regressive, since the golden age lay in the past. For Pericles, some centuries later, society is open, nature is malleable, and history is progressive. But by the end of that century, by the time of Euripides, society is seen as a weak illusion, nature a harsh reality, and history as meaningless.[43] The modern world has had its own cycles. At the dawn of modernity, Rousseau saw society as repressive, nature as good, and history as an illusion. Less than a hundred years later, Comte saw society as open, nature as malleable, and history as progressive. Today the cultural pessimists see society as a monster, nature as recalcitrant, and history as apocalyptic. Is this to be an endless recurrence?

The history of consciousness suggests a resolution. The uniqueness of man lies in his capacity, for self-consciousness and self-transcendence, to stand continually "outside" himself and to judge himself. This is the foundation of human freedom. It is this radical freedom which defines the glory and the plight of man. The modern view of man takes over only the aspect of freedom, not his finitude; it sees man as a creature of infinite power able to bend the world to his own will: Nothing is unknowable, Comte declared; Prometheus is my hero, Marx declared; Man can make himself, modern humanist psychology declares. It is man's incorrigible tendency towards self-aggrandizement, self-infinitization, and self-idolation which, in the political religions, becomes a moralizing absolute and, owing to the intrinsic egoism of human nature, masks a will to power.

Modern culture, particularly in its utopian versions, denies the biblical idea of sin. Sin derives from the fact that man as a limited and finite creature denies his finiteness and seeks to reach beyond it—beyond culture, beyond nature, beyond history. Evil, as Reinhold Niebuhr has put it, does not exist in nature, but in human history: ". . . human freedom breaks the limit of nature, upsetting its limited harmonies and giving a demonic dimension to its conflicts. There is therefore progress in human history; but it is a progress of all human potencies, both for good and evil."[44]

Thus there is a dual aspect to man as he stands recurrently at the juncture of nature and history. As a creature of nature, he is subject to its brutal contingencies; as a self-conscious spirit, he can stand outside both nature and history and strive to establish his own freedom, to control the direction of his fate. But human freedom is a paradox. Man is limited, subject to causal necessity, and bound to finite conditions; yet, because of his imagination, he is free to choose his own future and be responsible for his own actions. He is able to step over his own finiteness, yet that very step itself risks sin because of the temptations of idolatry—particularly of the will to power. That is the contradiction between finitude and freedom. That is the quandary of human existence.

FOOTNOTES

[1] My discussion of the Smithsonian legacy and its vicissitudes is taken from A. Hunter Dupree, *Science in the Federal Government* (Harvard University Press, Cambridge, Mass., 1957), Chapter IV, and Howard S. Miller, "Science and Private Agencies," in Van Tassel and Hall (eds.), *Science and Society in the United States* (Dorsey Press, Homewood, Ill., 1960), pp. 195–201.

[2] "Human history may be viewed as a process in which new needs are created as a result of material changes instituted to fulfill the old. According to Marx . . . the changes in the character and quality of human needs, including the means of gratifying them, is the keystone not merely to historical change but to the changes of human nature." Sidney Hook, *From Hegel to Marx* (University of Michigan Press, Ann Arbor, 1962), p. 277.

[3] Sidney Hook, "Materialism," *Encyclopedia of the Social Sciences*, Vol. X (Macmillan, New York, 1933), p. 219.

[4] "Introduction to the Critique of Political Economy." The essay, much of it in the form of notes, was intended as an introduction to the main work of Marx. The posthumous essay was first published by Karl Kautsky, Marx's literary executor, in *Neue Zeit*, the theoretical organ of the German Social Democratic Party, and published in English as an appendix to *A Contribution to the Critique of Political Economy*, Marx's work of 1859, by Charles H. Kerr, Chicago, 1904. The quotations in the text above are from pp. 310–12.

[5] Socialism has not come as the successor of capitalism. Communist China is technologically more backward than capitalist U.S.A. Is it socially more "progressive"? and on what dimensions: freedom, sexual styles, standard of living, communal care, personal dignity, social cohesion, attachment and loyalty to the country or party or leadership figure? Surely there is no way to "rank" these factors.

[6] As Webster's Second points out: "The conception of *nature* (Gr. *physis*, L. *natura*) has been confused by the mingling of three chief

meanings adopted with the word into English, viz.: (1) Creative or vital force. (2) Created being in its essential character; kind; sort. (3) Creation as a whole, esp. the physical universe. The main ambiguity is between *nature* as active or creative and *nature* as passive or created. In the original animistic view, the active vitalistic conception prevailed; but Plato sharply distinguished the passive material from the active formal element, and Aristotle continued the distinction in the conception of a moving cause, or God, as separate from the moved physical universe, or Nature. This antithesis is all but obliterated in pantheistic and naturalistic views. It appears in the pantheism of Spinoza, but the distinction of *natura naturans* and *natura naturata* serves only to discriminate two elements or aspects of the one organic being or substance. The two elements, in the forms of matter and energy, are retained in the modern physical or mechanical view, wherein nature appears as a material universe acting according to rules, but to all intents independent of God or purposive cause." (G. & C. Merriam, Springfield, Mass., 1955). p. 1631.

7 I am indebted for the illustration to John Wain, from a review of *Sexual Politics* in the London *Spectator*, April 10, 1971. As Mr. Wain writes: "Everyone agrees that happiness comes, and can only come, from living according to nature. And what is that? When woman is assigned a different role from man, is she being thwarted and twisted away from 'nature'? Or is it, on the contrary, the woman who wants to be treated exactly like a man who is turning her back on 'nature' and happiness?"

8 If nature is outside man, what does one do with the term *human nature?* Despite its ambiguities, it is probably indispensable. Yet, in the effort to keep my distinctions clear, I would use instead the term *human character*.

9 I realize that I am using the phrase "the history of nature" in a very different way from such physicists (or should one call him a natural philosopher) as C. F. von Weizsäcker, who asserts that nature is historic, since by history he means being *within* time, since all of nature itself is changing—and ten billion years ago there was neither sun nor earth nor any of the stars we know—and, following the theorem of the second law of thermodynamics, events in nature are fundamentally irreversible and incapable of repetition. My history of nature, here, is within the time frame and conceptual map of nature's transformation at the hands of man, and the understandings of nature by man. See C. F. von Weizsäcker, *The History of Nature* (University of Chicago Press, Chicago, 1949).

10 "Protagoras," in *Plato: The Collected Dialogues,* edited by Edith

Hamilton and Huntington Cairns, translated by W. K. C. Guthrie (Bollingen Series LXXI, New York, 1966), lines 320d to 322, pp. 318–19.

11 Aeschylus, *Prometheus Bound*, translated by Edith Hamilton, in *Greek Plays in Modern Translation*, edited by Dudley Fitts (The Dial Press, New York, 1953), pp. 508–9, 519–20.

12 In this section I have drawn primarily from E. J. Dijksterhuis, *The Mechanization of the World Picture* (Oxford University Press, London, 1961); Charles C. Gillespie, *The Edge of Objectivity* (Princeton University Press, Princeton, N.J., 1960); Arthur Koestler, *The Sleepwalkers* (Hutchinson, London, 1959); John Herman Randall, Jr., *The Making of the Modern Mind* (Houghton Mifflin, Boston, 1926), especially for the quotations from Descartes and Spinoza; and Joseph Mazzeo, *Renaissance and Revolution* (Pantheon Books, New York, 1965), especially on Galileo. Unless otherwise noted, the quotations from Descartes and Spinoza are taken from Randall.

13 As Descartes wrote:

. . . instead of that speculative philosophy which is taught in the schools, we may find a practical philosophy . . . by means of which, knowing the force and action of fire, water, air, the stars, heavens and all other bodies that environ us, as distinctly as we know the different crafts of our artisans, we can in the same way employ them in all those uses to which they are adapted, and thus render ourselves the masters and possessors of nature.

The Philosophical Works of Descartes, trans. by Elizabeth S. Haldane and G. R. T. Ross (Cambridge University Press, London, 1931), Vol. I, p. 119.

14 For Lynn White's view, see "The Historical Roots of Our Ecological Crisis," in *Machina ex Deo* (M.I.T. Press, Cambridge, Mass., 1968), esp. p. 90. "For nearly two millennia Christian missionaries have been chopping down sacred groves, which are idolatrous because they assume spirit in nature." For a different view see Clarence J. Glacken, *Traces on the Rhodian Shore* (University of California Press, Berkeley and Los Angeles, 1967), Chapter 4, esp. p. 162.

15 See Edgar Zilsel, "Problems of Empiricism," in *The Development of Rationalism and Empiricism*, International Encyclopedia of Unified Science, Volume II, Number 8 (University of Chicago Press, Chicago, Ill., 1941); Dijksterhuis, *The Mechanization of the World Picture, supra*, pp. 241–47; and, as a case study, Joan Gadol, *Leon Battista Alberti* (University of Chicago Press, Chicago, Ill., 1969).

16 Edgar Zilsel, "The Genesis of the Concept of Scientific Progress," in

Philip Wiener and Aaron Noland (eds.), *Roots of Scientific Thought* (Basic Books, New York, 1960), p. 251.

[17] One might take this activity as an illustration of the range of interest of the man of letters, before "the two cultures." As Franklin Ford observed even of culture a century later:

Through most of the eighteenth century, a truly learned man, wherever his most highly developed competence might lie, claimed and was conceded the right to move with interest and confidence over a range which we should now describe as covering physical sciences, social sciences, and humanities. What was Montesquieu—political scientist, sociologist, historian, aesthetic philosopher, critic of morals? Before selecting a label, we might reflect that his earliest scholarly prize was won with a paper concerning nervous and muscular reactions observed in the thawing of a frozen sheep's tongue.

"Culture and Communication," unpublished paper for the American Academy of Arts and Sciences conference on *Science and Culture*, Boston, May 10–11, 1963.

[18] *The Edge of Objectivity, supra*, p. 79.

[19] As Charles Gillespie has observed: "His was the philosophy that inspired science as an activity, a movement carried on in public and of concern to the public. This aspect of science scarcely existed before the seventeenth century. . . . This is comfortable democratic doctrine, and it is obvious why Baconianism has always held a special appeal as the way of science in societies which develop a vocation for the betterment of man's estate, and which confide not in aristocracies, whether of birth or brains, but in a wisdom to be elicited from common pursuits—in seventeenth-century England, in eighteenth-century France, in nineteenth-century America, amongst Marxists of all countries." Ibid., p. 75.

[20] Francis Bacon, *New Atlantis*, in *Famous Utopias*, edited by Charles M. Andrews (Tudor Publishing Co., New York, n.d.), pp. 235–72, esp. 263, 269–71.

[21] Quoted in Alfred Schmidt, *The Concept of Nature in Marx* (NLB, London, 1971), pp. 129–30.

[22] Ibid., pp. 131–32.

[23] *Capital*, Volume III, pp. 799–800. There is a subtle difference here between Marx and Engels which is more than of interest to the specialist. In *Anti-Dühring*, which is better known for its formulation of the "leap to freedom" than the passage in *Capital*, Engels assumes that when man "for the first time becomes the real conscious master of nature, because and in so far as he has become master of his own social organization," all necessity, i.e., *all* labor, is abolished.

But Marx was less utopian and argued that some necessity would always remain. This question is explored by Alfred Schmidt, op. cit., pp. 134–36.

24 D. Ryazanoff (ed.), *The Communist Manifesto of Karl Marx and Friedrich Engels* (Russell & Russell, New York, 1963), pp. 29, 31–32. And there was still more to come. Wilhelm Liebknecht, one of the founders of the German Social Democratic Party, tells of his conversation with Marx when he joined the German Socialist Workers Club in London in 1850: "Soon we were on the field of Natural Science, and Marx ridiculed the victorious reaction in Europe that fancied it had smothered the revolution and did not suspect that Natural Science was preparing a new revolution. That King Steam who had revolutionized the world in the last century had ceased to rule, and that into his place a far greater revolutionist would step, the electric spark. And now Marx, all flushed and excited, told me that during the last few days the model of an electric engine drawing a railroad train was on exhibition in Regent street. 'Now the problem is solved—the consequences are indefinable. In the wake of the economic revolution the political must necessarily follow. . . .'" Wilhelm Liebknecht, *Karl Marx: Biographical Memoirs* (Charles H. Kerr, Chicago, 1901), p. 57.

25 Can there ever be an end to scarcity? What is striking is that Marx, like all nineteenth-century thinkers, conceived of abundance as a plethora of goods. But, in contemporary society, there are many new—and crucial—kinds of scarcity which were never envisaged, notably the scarcity of time. And if we think of abundance, as an economist does, as items being a "free good" (i.e., as having a zero cost, as clean air once had), then one finds in contemporary society that a host of new activities, such as information, or the coordination of activities in participatory situations, create rising costs and therefore become increasingly scarce resources. For an elaboration of this argument, see the section "The End of Scarcity" in my book *The Coming of Post-Industrial Society* (Basic Books, New York, 1973).

26 *Techne*, it should be noted, though we use it to denote technology, in Greek means art, both in the sense of a craft used by a craftsman (artisan), and as Aristotle defined it, as consisting "in the conception of the result to be produced before its realization in the material." ("De Partibus Animalium," 641, in *The Basic Works of Aristotle*, ed. by Richard McKeon (Random House, New York, 1941), pp. 647–48). I am indebted to Emanuel Mesthene for the suggestion.

27 I owe this definition to Harvey Brooks.

The Frontiers of Knowledge

28 Mervyn Jones, "Tomorrow Is Yesterday," *New Statesman*, Oct. 20, 1972.

29 The extraordinary increase in world-wide demand in the past decade has led to new shortages in food, materials, and energy. That these new shortages lead to higher prices goes without saying. Whether these shortages will persist, i.e., that no new sources of materials, energy, or intensive increase in food yields will be found, is moot. But the major point remains: it is the emphasis on demand, an economic fact only a century old, that has led to this state of affairs.

30 *On Growth and Form* (Cambridge University Press, London, 1963) (1st edition, 1917), p. 27. Taking the story of Jack the Giant Killer, Thompson pointed out that Jack had nothing to fear from the Giant. If the Giant were ten times as tall as a man but built like one, he was a physical impossibility. According to the square-cube law, the Giant's mass would be 10^3, or a thousand times Jack's, because he was ten times as big in every dimension. However, the cross section of his leg bones, if he had the shape of a man, would have increased only in two dimensions, 10^2, or a hundred times as big as Jack. A human bone will not support ten times its normal load, and if the Giant tried to walk he would break his legs. Jack was perfectly safe.

31 "Can We Survive Technology?" *Fortune*, June 1955, reprinted in *The Fabulous Future* (Dutton, New York, 1956), p. 34.

32 See, for example, D. H. Meadows et al., *The Limits to Growth* (New American Library, New York, 1972), and, for a telling critique, Carl Kaysen, "The Computer That Printed Out W*O*L*F," *Foreign Affairs*, July 1972.

33 I take these figures from an essay by John Platt, "What We Must Do," *Science*, Vol. 166, Nov. 28, 1969, pp. 1115–21.

34 I take these examples from Martin Shubik, "Information, Rationality and Free Choice," in Daniel Bell (ed.), *Toward the Year 2000* (Houghton Mifflin, Boston, 1968); and Robert A. Dahl, "The City in the Future of Democracy," *American Political Science Review*, December 1967.

35 The temper is not restricted to ecologists. Alfred Kazin cites the titles of some recent cultural-social analyses of "our situation": *Reflections on a Sinking Ship*, *Waiting for the End*, *The Fire Next Time*, *The Economy of Death*, *The Sense of an Ending*, *On the Edge of History*, *Thinking About the Unthinkable*.

36 Theodore Roszak, for example, writes: ". . . we must not ignore the fact that there *is* a natural environment—the world of wind and wave, beast and flower, sun and stars—and that preindustrial peo-

76

ple lived for millennia in close company with that world, striving to harmonize the things and thoughts of their own making with its non-human forces. Circadian and seasonal rhythms were the first clock people knew, and it was by co-ordinating these fluid organic cycles with their own physiological tempos that they timed their activities. What they ate, they had killed or cultivated with their own hands, staining them with the blood or dirt of their effort. They learned from the flora and fauna of their surroundings, conversed with them, worshiped them, and sacrificed to them. They were convinced that their fate was bound up intimately with these non-human friends and foes, and in their culture they made place for them, honoring their ways."

What is striking in this evocation of a pagan idyl is the complete neglect of the diseases which wasted most "natural" men, the high infant mortality, the painful, frequent childbirths which debilitated the women, and the recurrent shortages of food and the inadequacies of shelter which made life nasty, brutish, and short.

37 Jacques Ellul, *The Technological Society* (Knopf, New York, 1964), Chapter II, *passim*. What is striking in this unsparing attack on technique is Ellul's omission of any discussion of nature, or how man must live without technique. (The word *nature* does not appear in the index, and there are only a few passing references to the natural world, e.g. p. 79.) As Ellul's translator, John Wilkinson, writes in the Introduction: "In view of the fact that Ellul continually apostrophizes technique as 'unnatural' (except when he calls it the 'new nature'), it might be thought surprising that he has no fixed conception of nature or the natural. The best answer seems to be that he considers 'natural' (in the good sense) *any* environment able to satisfy man's material needs, *if* it leaves him free to use it as means to achieve his individual internally generated ends." Ibid., p. xix.

38 Ibid., p. 97.

39 The distinction is made in the report of the National Academy of Sciences, *Technology: Processes of Assessment and Choice*, published by the Committee on Science and Astronautics, U. S. House of Representatives, July 1969. See p. 16.

40 The traditional story is told that an impatient man once asked Rabbi Hillel to tell him all there was in Julaism while standing on one foot. The Rabbi pondered, and replied: "Do *not* do unto others as you would *not* have them do unto you. All the rest is exegesis."

41 Protagoras, op. cit., sections 322 c–d, pp. 319–20.

42 James Redfield, "The Sense of Crisis," in John R. Platt (ed.), *New Views of the Nature of Man* (University of Chicago Press, Chicago, Ill., 1965), p. 122.

[43] Ibid., pp. 128, 135, 142.
[44] Reinhold Niebuhr, *The Nature and Destiny of Man* (Charles' Scribner's Sons, New York, 1945).

HISTORY, TECHNOLOGY, AND THE PURSUIT OF HAPPINESS

EDMUNDO O'GORMAN *Mexican historian Edmundo O'Gorman teaches the philosophy of history and colonial Mexican history at the National University of Mexico, of which he is Professor Emeritus.*

In 1958, in a series of lectures on "The Invention of America" (published in 1961 by the Indiana University Press), O'Gorman joined philosophy and history in a subtle and original study of sixteenth-century Europe's bafflement at the knowledge of the existence of an unsuspected continent. Throughout a long and varied career, his special interest has been the history of the exploration and settlement of Spanish America.

Writer of numerous books, his most recent LA SUPERVIVENCIA POLÍTICA NOVO-HISPANA *(1969), O'Gorman also has edited the works of Joseph de Acosta (*HISTORIA NATURAL Y MORAL DE LAS INDIAS, *1962), Bartolomé de Las Casas (*APOLOGÉTICA HISTORIA, *1967), and Toribio Motolinío (*MEMORIALES, *1971).*

He has lectured at numerous institutions in the United States, and spent a one-year term as visiting professor at Brown University.

I

IT IS FITTING that at the outset of my talk I should express my thanks to the distinguished historian and director of the National Museum of History and Technology, for having invited me to take part in these lectures sponsored by Doubleday & Company to mark its seventy-fifth year in publishing and to honor the memory of its founder, Frank Nelson Doubleday. It is no less fitting, therefore, to congratulate this highly respected publishing house for the anniversary and to wish that it may enjoy for many years to come a long and productive life.

Within the general program of these lectures, designed to explore the impact of technology on various fields of human endeavor, I have been asked to examine the role it has played in history. The complexity of such a formidable theme will be apparent to everyone, and it should not be surprising that in the brief space of a single lecture many aspects will perforce be left untouched and many strings dangling. And also, inevitably, it will be impossible to justify fully certain notions which must run the risk of appearing dogmatic or even as arbitrary assertions. Such deficiencies and danger did not weigh enough, however, to deter me from the attempt, if and when I may rest, as I very much hope, on the benevolence of my audience. It hardly seems necessary to warn that the subject of our concern involves nothing less than a vision of universal history and that, therefore, a previous notion about man must be basic to our inquiry, a notion which envisages man as an entity projected towards the future, responsible for his own being and, as we shall see, inexorably engaged in technological pursuits. I gladly acknowledge my debt to the many eminent historians and philosophers beginning with Dilthey—to only mention a modern—who have forged such a fundamental notion to apprehend historical reality and the meaning of its happening. But in this connection it would be unfair not to specially remember the late Spanish philosopher José Ortega y Gasset, whose essay "Meditation on Technique" (*Revista de Occidente*, Madrid, 1939) has been particularly valuable.

II

There was a time, recent in years but historically remote, when intellectuals and artists held everything connected to technology as alien to culture. A man of culture was not and did not wish to be concerned with that dreary world of engineers, industrialists, and factory hands and only took note of its existence as some inferior sphere, necessary perhaps, but severed and even hostile to that heavenly circle, the habitat of the beautiful people and exquisite minds devoted to things spiritual.

The spokesmen for that assumed infra-orb reacted, for their part, with a parallel scorn towards those given over to books and to the cultivation of the higher arts, holding them in ridicule as a group of parasites who, masquerading behind an impenetrable barricade of tastes forbidden to common mortals and of an esoteric prattle, contributed in no way to mankind's progress and welfare.

This conflict between an outworn romanticism and a vulgar materialism predominated—explainably—in the United States at the time when American society was overrun by the most powerful industrial expansion to which history bears witness. It became a handy subject for satiric caricature and comic strips to depict the constant friction in upstart marriages, in which the woman, self-appointed priestess of artistic refinement and gentility, benefited from but slighted the activities of her husband, who, while a bold and farsighted adventurer in business and industry, exhausted his own spiritual cravings in the delights of the poker table or the baseball field.

Today, with nuclear weapons and space odysseys, nothing is left of that ridiculous situation except a vestige, on the one hand, in the jejune snobbism of a few interior decorators and, on the other hand, in the campaign perorations of a certain breed of politicians. But it is important to underline the legacy, because not infrequently technical activities are still disassociated from cultural pursuits, as if they belonged to respectively independent historical provinces. No one doubts any more the vast importance of technology, and even its contribution to the arts is widely accepted. Yet, ultimately, technology is viewed with deep suspicion and is envisaged as something susceptible of being considered separately from the true needs of human welfare. As something, therefore, which has had the greatest influence on history, shaping its course and marking it with indelible features, yet as something which merely *happened* to man, that is to say, as not essential or constitutive to his being. Technical activity is thus conceived of as an historical *accident*, a happening which happened to happen but which might not have happened at all. This misunderstanding of the true nature of technology is clearly evident in the touching but vain hopes of those who, believing it to be a sort of extraneous telluric force, would flee its evils to some pastoral refuge in perfect harmony with nature. It is that, indeed, which invites so many of the young today to embrace a so-called primitive life which at the end of a year or so will have victimized them with disease, discord, and unbearable tedium. Who hasn't undergone the appalling torture of a truly primitive picnic? The Victorians knew better: they held their picnics on the lawn in the cool shade of bygone medieval grandeur and thought of them as fitting occasions to read Horace and sip champagne.

III

In vigorous opposition to the notion of technique as incidental to man's life, Oswald Spengler[1] offers a firm starting point. Technique, he tells us, is universal to all forms of life. It may not be ascribed to any one given period or species. It is, in essence, the assertion of all living entities in and as opposed to the environment or, if you wish, to nature. The existence of any form of life implies technique. This fundamental notion destroys the all too general supposition that technique is only concerned with the manufacture and use of tools and machines and discovers it to be the means by which all living beings are maintained in the natural circumstances in which they find themselves or, to say it in Spengler's words, it consists in the tactics employed by life to realize itself. The end pursued by technology, therefore, is one and the same as that pursued by life.

Such an omnifarious notion includes the whole biological scale from the amoeba to man, considering that all living beings have a particular way of behaving in the struggle which is living. But, clearly, as there are disparate forms and hierarchies of life, there also are different ways and refinements in technology. In comparison to the most sophisticated techniques of botanical life—the way certain plants reproduce, for instance—animal technique is ever so much richer and more diversified, and here one may distinguish between passive technique—the camouflage of the chameleon, the flight of the gazelle—and the strategy and craftiness of the big predatory animals. Both imply a different way of life; the one seeks invulnerability, the other takes dangerous risks. Spengler places man among the predatory animals, and from this he derives his idea of history as a process prompted by the will to conquer and believes that human technology is the answer to that will, fully developed only, however, in Faustian civilization, exclusive to nations of Germanic stock.

Undoubtedly, Spengler's powerful and alarming vision contains some truth, yet not enough to make us follow in its wake: after all, man surely is more than just a rapacious beast. But before turning from the tracks of this erstwhile most renowned historian-philosopher we may take advantage of one of his more apposite distinctions. He points out that all animal technique, whether passive or aggressive, possesses a common trait which sets it back at a gigantic distance from human technology—namely, that it does not introduce significant changes in the environment and, in any case, there is a void of consciousness as to the tremendous meaning attached to such an extraordinary effect. The life of an animal is like a toy in nature's hands and it remains inwardly identical, or, as Spengler says, in the interior recess of the soul. It is because of this narrow limitation that we can say with truth that animals live by instinct, which, of course, does not preclude intelligence, audacity, purpose, and admirable accomplishments.

Such a profound dissimilarity between animal and man

warrants the exclusive usage of the word technology to human activities, and from here on we shall thus employ it.

IV

Following Spengler's suggestion, we have asserted that the introduction of change in the environment is the specific trait of technology. Let us take a close look at this peculiarity, guided by Ortega's insight and authority.[2]

Man is not self-sufficient. In order to live, he must gratify his needs. All other forms of life are subject to the same predicament, but the very remarkable difference is that while other forms submit to the circumstances in which they find themselves—an acquiescence which makes animals not to feel, strictly speaking, their needs as such—man alone is conscious of nature's hostility to the fulfillment of his wishes, and, therefore, feels the constraints of nature as something imposed on him, a sort of injustice done to him and against which he rebels. Whatever it is he does not find at hand, awakens a state of mind and becomes thus subjective in such a way that it is felt as something which he lacks. In a word, that lack is conceived *sub specie* of necessity. Like the animals, man is not self-sufficient, but the colossal difference is that he is conscious of that tragic condition.

As we go along, the incalculable consequences of such a peculiarity will become apparent. Let us fix our attention for the moment, on the most immediate. Indeed, when it so happens that nature provides, both man and animal simply and directly gratify their needs. But when that is not the case—which, as just explained, is what generates the feeling of necessity—man is moved to a very different kind of activity: it no longer consists in merely satisfying his want, for he must previously procure the means to do so. Obviously, heating oneself, for instance, is far from being the same thing as having to build a fire in order to attain that end. We can clearly see, then, that this other kind of prior activity—exclusive to man—involves inventing processes and the manipulation of things for the purpose of having within reach at all times the means to satisfy whatever may have been felt as a need.

We may say, then, that the activity displayed by man in response to his feeling of a need resolves itself by imposing on the environment certain conditions or improvements which nature lacks. Such activity really means, therefore, that in order to live and fulfill the possibilities of his being, *man amends or, better, reforms nature in such a way that he may satisfy his wants.* And this is what technology is all about.[3]

V

Let us take a closer look at our definition of technical activity. The first thing about it is that, like any other activity, it may be considered from the objective or subjective point of view, and our immediate concern will be to study in their turn these two aspects of the question.

It is obvious that, in saying that man amends or reforms nature, we are objectively saying that he introduces changes of the most varied kind, such as, for instance, altering the natural distribution of wooded land by razing forests; creating innumerable man-made things that are not naturally produced; or by inventing, say, with what are only vibrations in the air, what we know as music. In other words, by means of his technical activity man manufactures a special and new environment, an artificial nature different from although supported by the original, pristine nature.

In the remote dawn of man's history this other nature was hardly visible, because few changes in the environment were needed to satisfy primitive biological needs and spiritual wants: a footpath in the forest, a small ditch, rudimentary tools and domestic objects, rupestrian inscriptions and clay or wooden images to appease the wrath of the mysterious and menacing forces which beset man on all sides. But in essence those modest marks of man's technical activity and ingenuity have the same meaning as the colossal transformation that modern technology has imposed and will further impose on the face of the earth. Each transformation, the great and the small, gives witness to man's constructive power in changing the natural environment so that he may live in it according to his needs or wishes, because, as we shall see, it comes to the same thing.

But of the greatest importance, we must emphasize, is the dramatic dualism implied in what we have explained. Indeed, man finds himself submerged in an environment which, though usually thought of as one unit, is really two coexisting entities, distinct both in extension and kind. To acquaint ourselves with their peculiar features it is advisable to identify them separately by their proper names: by *universe* we refer to unchanged natural reality, whereas we shall reserve the name of *world* to signify artificial, or man-made, nature.

The concept of universe is inclusive, by definition, of all that exists. The universe is not man-made, and because of that, and in that sense, it does not belong to him. It belongs to God, whether conceived in fetishist, mythological, theological, metaphysical, or scientific terms, but always as the power responsible for reality.

On the other hand, the world is man-made, and because

of that, and in that sense, it is his; it belongs to him, not to the Godhead, who has not and cannot have any use for it. The world is mankind's possession and tenure, the result of the imagination, labor, courage, and endurance of countless generations since the earliest age, at the time when man became man.[4]

Now, this duality is evident in two ways. The first refers to the extension, the second to the class or kind of the one and the other entities. Indeed, it is not difficult to see that the world, being finite, is lodged within the universe, like, let us say, a capsule surrounded on all sides by the infinite space of the universe; but also, the world being artificial, such a capsule is a sort of tumor in the immaculate body of the universe.[5] These two traits that traditionally have kept apart the idea of the universe from the idea of the world must be borne in mind, because we are to come back to them when, later on, we will try to explain the vast difference between contemporary scientific technology and that of earlier ages. For the present, it will suffice to know that the reform introduced in nature by man does not merely mean the bringing about of changes but, much more radically, a confrontation with another kind of reality, which—it may be surmised—will eventually end up by swallowing it entirely.

VI

Having considered technical activity on its objective side, we are to look at it from the subjective point of view by inquiring after its consequences, if any, to man's being.

If, as has been explained, man rebels against adverse conditions of his environment and modifies it to suit his wants, even a cursory view of this action will show that such a lack of conformity does not really concern nature as such, but the kind of life allowed to man on her terms, which is not the same thing. The distinction will be clearer if we put it by saying that for man it is not enough to maintain himself in nature—as is the case with other living entities—but he must maintain himself in a better way than nature will allow. Man does not, purely and simply, want *to be* in nature; he wants to shape it; he wants and seeks his own *well-being.* Human life is not only a struggle for life; it is a struggle for a better life,[6] and precisely because that is what man aspires to, he is moved to change the environment and reform nature in accordance to his desire.

But here we must caution against the common misunderstanding of a better life as consisting of the possession of means in greater quantity or superior quality—or both. This is a fallacy. In the first place because there are many to whom a better life means exactly the opposite, but in the second place, and more fundamentally, because greater number or superior quality of means do not merely change the quality of life; they change life itself by opening the possibility of a different life, or, as is usually and aptly said, of a "new life." The fallacy, therefore, lies in assuming that life is something previous to and beyond the act of living, and the absurd sequitur will be that life is not the living of it, but a series of happenings that concern it but do not in the least affect it. This, however, is not all, for it must be noted that the desire for a better life not only implies, as we have just seen, a different life but also the aspiration of being other than the way in which one is. Such a consequence is obvious, as will appear in a couple of instances. If I desire a different life, let us say an immortal life, it is really because I aspire *to be* immortal; if I wish to lead a gentle life, it is that I wish *to be* a gentle man. And here, suddenly, we have uncovered man's extraordinary and unique peculiarity: he is an entity capable of ceasing to be what he is at a given moment in order to be in a different way. We may thus understand that man's being is not—as tradition has it at least since Parmenides—an essence, that is to say, an unalterable, immutable substance like, for instance, that of a stone or a star. Man is not a *thing;* he is, according to Montaigne's dictum "undulating and mutable"; an entity where being is not static, but dynamic; an entity projected towards the future in an ever-changing process which, as far as we know, only stops with death.[7]

We may now return to the chief point of our inquiry, and in the light of the foregoing explanations it will appear that man's lack of conformity with the kind of life that nature itself will grant him and the correlative wish for a better life are at bottom a lack of conformity with nothing less than with *what he is.* In that deep-rooted rebellion, therefore, we find the secret spring that incites and moves all technical activity, which thus discloses itself as the means in man's power to be what he desires to be. But then, the world—that artificial nature manufactured by man's technical activity—discloses itself, in turn, as the environment required by man to become what he wishes to become; the only environment, therefore, where he may actually realize himself.

VII

But what is it that man wishes to be? Here, at last, is the crucial question. Up to this point we have only hinted at the answer when asserting that man wishes his well-being or that he desires a better life. Now, no one will object, I believe, that those assertions are different ways of saying the same thing—namely, that man aspires to be happy. We may then summarize our whole argument by saying that man's necessity to be happy is the fundamental impulse in human life and that this prompts all the other needs. It is the necessity of necessities.

This answer to our crucial question may be regarded as supremely unreliable, because few concepts—if any—are as relative or subjective as that of happiness. Furthermore, nothing is more unstable and fickle, as we all know from personal and painful experience. So with reason we may be told that our castle is built on moving sand.

There is, however, no cause for despondency, provided we remember that in answering our questions we did not speak of *feeling* happy, but of *being* happy. The former refers to a contingent possibility, the latter to a permanent state. Let us take two examples to make our point. It will be granted, I assume, that the happiness felt by a woman in seeing the man she loves—but which, on his parting, will turn to misery—is not the same as the happiness of a monk who covets nothing the world can offer. Wherein lies the difference? Undoubtedly in the fact that the woman needs to see the man she loves, whereas the monk has no needs at all, and that is why we say about him, not that he *feels* happy, but that he *is* happy.

We may now understand that man's aspiration to be happy is his desire to be exempt from want or, to say it more technically, to become a *non-necessitated entity*.[8]

This is, then, the decisive notion for a thorough comprehension of the ultimate objective of technology. Indeed, now we may see that the artificial reality which technology imposes upon natural reality—the man-made world—is the attempt to bring about the requirements of an environment in which man may reach the condition of a non-necessitated entity: a sort of heaven on earth where the state of beatitude promised by religion after death will be possible during mortal life.[9] Another thing, of course, is whether man will succeed in such an enterprise or in the trying will destroy himself. We may have time to ponder this alternative after having considered technology, as we now understand it, within the process of historical events.

VIII

If technical activity is moved by man's desire to be what he wishes to be (we now know what that is) this implies the capacity of being able to imagine previously the new life or way of being desired for the future. Lacking that capacity, the desire to be one way or another could not arise and neither would the impulse to reform nature. Man's life would be a repetitious process like that pertaining to other forms of life. In other words, man would not be human. Man's supreme faculty, consequently, is not reason; it is imagination, the truly "divine spark" that makes sense of the ancient mythological intuition of man having been made in God's likeness. Indeed, imagination conceives reality before it exists, a prodigious achievement, second only to the act of creating *ex nihilo*, reserved to the Godhead.

Logically prior to all technical activity, we find therefore an act of the imagination by virtue of which man visualizes in anticipation what it is he wants to be and conceives a program thereof. Such a program may well be called the "project of life," the proper understanding of which entails three points.

In the first place, it is not an abstraction. The project of life contains a concrete program, such as, for instance, those which have given birth to the ancient Eastern and New World civilizations, Greek-Roman antiquity, Christianity and Buddhism, and modern man of Western cul-

ture. This trait is all-important to grasp the function of a
project of life as the guiding principle of technical activity,
because it determines its kind and its concrete objectives.
Some societies will erect pyramids while others will build
temples in response to the way in which they conceive
their gods. The wealth of forms and styles in art, the
differences in institutions and social habits, and, in other
words, all the variegated prospects offered on the stage of
the great theater of universal history are but the visible
forms of different ways in which man has imagined his life
and destiny: the echoes of the diversity and profusion in
the projects of life devised by man ever since he became
worthy of that name.

In the second place, notwithstanding the variety of
concrete historical projects of life, they all recognize a fundamental unity. The reason for this is that the supreme objective is always the same one, namely—as we already know
—the attainment of happiness. Such a purpose, therefore,
defines the project of life of all the historically possible
concrete projects of life.

In the third place, the above being the case, one may ask
which—if any—of the concrete historical projects that
have been tried may be regarded as the most successful.
Now, the answer to this question cannot be given unless
we first determine the two requisites of what would be the
ideal project of life imaginable.

The first requisite is that the happiness which is the objective pursued should be attainable in this life. That not
being the case, the project in question would not be a project of life, but a project of death.[10]

The second requirement is that the project of life in
question should prompt a change in the environment so
that it may no longer oppose man's wishes for a better life.
The corresponding technology of such a project would

93

have to reform nature in such a way that (a) it would include all of it, and (b) that it would control and dominate it. Under those conditions, indeed, man would not only cease to suffer want, but the causes of such a possibility would have disappeared.

It may be surmised that, in describing the above conditions, we have had in mind the project of life of our modern Western Euro-American civilization, and it is now for us to show that it actually fulfills them.

IX

As to the first requirement, a few words will suffice. It seems indisputable that our modern civilization is grounded on a program which envisions life here and now, and although Christianity postulates a world beyond the grave, such a belief has never been a serious obstacle to the pursuit of happiness in this life, but pre-eminently so since the Renaissance, when modern civilization has its true beginning. Pascal's famous argument is all too significant: it is a good gamble—so it runs—to believe in an afterlife, because if it does not exist there can be no loss, whereas if it exists, we may well lose all.

The second requisite calls for a much more detailed explanation. It will be remembered that it has two aspects, which we shall consider separately, the first being that the reform imposed by technology on nature must include the whole of it. Let us turn, then, to history in search of the moment when an event of such magnitude took place. We need not look long for it—it happened at the end of the fifteenth and the first decades of the sixteenth centuries and is misleadingly known as the discovery of America. Some ten years ago I attempted to disclose in a small book called

The Invention of America[11] the profound spiritual revolution brought about by that most singular and memorable event, and here a condensed abstract must suffice.

The world at the time Columbus crossed the Ocean was conceived as an entity formed of three parts: Europe, Asia, and Africa. These were thought of as intrinsically different entities, and for physical and theological reasons it was impossible to admit the existence of other parts. The threefold division did not, therefore, constitute an arithmetical series; it was a closed hierarchical series. Consequently, the world's structure was definitively and permanently made, and nothing could change it. It occupied only a portion of the globe called the Island of the World, surrounded by the waters of the Ocean. The Ocean, as well as the rest of the terrestrial globe, belonged to cosmic space, a part, therefore, of the universe, belonging to God. In this ancient conception, the world was a small province, lodged within the infinity of the universe, and in which God graciously allowed man to live, a sort of prison that did not even belong to man, since he had not made it. Man thus turned out to be a serf infinitely thankful for his prisonlike cosmic habitat.

Subsequent explorations made it experimentally clear that the new-found lands could not be a part of Asia and, consequently, had no place in the threefold and closed structure of the world. The alarming result was that those lands had to be conceived as a "fourth part" of the world, and were given the name of America.

Now, the all-important point we want to make here is that the obliged acceptance of a fourth part of the world not only brought about the collapse of the old way of conceiving it, but that it prompted a new way in which the world suddenly overflowed its ancient barriers and embraced the whole terrestrial globe, as a consequence of the

Ocean having lost its old status to become a mere geographical accident. But this process did not stop there, because—*nota bene*—the existence of a fourth part of the world necessarily implied the real possibility of the existence of a fifth, a sixth, a seventh part, and so on *ad infinitum*. In other words, the ancient closed, three-part, hierarchical series became an arithmetical series, that is, an unlimited series of parts of identical nature. The world, therefore, really ceased to be conceived as made of "parts" and, consequently, embraced the whole universe—there being no longer any way to distinguish one part from another.[12]

Such, then, is the great cultural revolution brought about by the oceanic explorations carried on during the late-fifteenth and early-sixteenth centuries and which—in a very real sense—prepared the ground for that other, better-known spiritual adventure called the Copernican revolution. Thanks to the audacity and endurance of those explorers and to the courage of the men who challenged the validity of the time-honored threefold structure of the world, the duality—in one of its aspects—between the universe and the world (cf. V, *supra*) was finally overcome as man broke loose from his ancient cosmic prison and serfdom.

X

Having shown when and in what way the world extended its realm so as to include the whole universe, I must now show when and how Western man took possession of it and made it his own.

This stupendous feat was definitively achieved at a time contemporaneous or nearly—not casually, of course—with the revolution we have just described. Indeed, throughout the millenniums up to the Renaissance, technology employed direct and basically elementary means, so the reforms man imposed on nature left her, so to say, virginal and internally intact. On the whole, technical activities did not go beyond the extraction and exploitation of raw materials, the change of external features of the original landscape, and the subjugation of plants, animals, and man himself. Muscle was essentially the sole source of energy. Now, such methods—without disparagement for the many and extraordinary results achieved—entail, more than a true domination over nature, mere exploitation and utilization of natural resources and raw material. In pointing this out we are, of course, referring to the gigantic difference in regards to modern technology. Indeed, we are all aware that there was nothing like it before the discovery, during the fifteenth and sixteenth centuries, of physical-mathematical science, which opened the road, no longer to merely utilizing this or that natural resource or to solving such and such a technical problem, but to harness, for any task or purpose, nothing less than the incommensurate energy contained in matter, the energy, therefore, which sustains universal reality.

From that time on, the universe ceased to be a mysterious entity endowed with secret qualities—virtues and essences—governed by God through the inscrutable intentions of His providence. It became a vast system of foreseeable relations between phenomena, capable of being stated in mathematical language and controlled by mechanical means. I have referred, obviously, to the invention of the machine, a contrivance of—in principle—unlimited power, endurance, and productivity, like the universe itself, which, indeed, was from then on conceived as nothing more than a gigantic and perfect machine, the mover of all

97

machines and, consequently, the machine of all possible imaginable machines.[13]

In this case, as in the previous one, the universe was also invaded by man's made world, this time not merely in extension but in its very core. Thus the ancient duality—in its second aspect—between artificial and original reality (cf. V, *supra*) was overcome.

We may, then, state the obvious conclusion: namely that, as surmised, Western technology, alone in historical experience, complies with the requirements and conditions already mentioned (cf. VIII, *supra*) as indispensable to establish and build a world in which man may actually reach his supreme calling by achieving happiness in this mortal, fleeting life.

But to this conclusion we must add that modern scientific technology is not only pre-eminent, but the only one capable of attaining such a level of excellence. Clearly, man can go no further in that direction, as there is no other universe which he can swallow. And that the *ne plus ultra* has been reached may be confirmed by pondering the evident historical fact that Western man's project of life has been universally accepted as the only possible one with a future. In our day there are no other civilizations besides our own, since one can hardly count as such the poor relics of different ways of life which have survived but are inevitably doomed to disappear or live on only, if lucky, in the exhibit halls of the Smithsonian Institution. No one will contest, for instance, that Mao's China with its atomic bomb and its Western Marxist doctrine is anything else but a late and conspicuous daughter of a culture which out of habit we still call Western but should designate as universal. And where the extremity of the situation brought about by the uniqueness and excellence of scientific technology may be strikingly and dramatically evidenced is in that it has put humanity in the very real possibility of self-destruction. The enormity of such a circumstance should suffice to make us aware that since the inception of scientific know-how, history entered an intrinsically new phase in its perhaps tragic course.[14]

Without undue immodesty, I believe that by having shown what seems to me to be the true nature of all technical activity, and its fundamental purpose of establishing the necessary conditions for man to achieve happiness in this life, and by pointing out why scientific technology is the one and only way to reach that supreme goal, I have acquitted myself—to the best of my ability—of the task I undertook by accepting the distinction of partaking in this series of lectures. Here, therefore, I could put an end to this talk; it does seem appropriate, however, to say a few words concerning what appears to be the basic alternative for the future.

XI

What we have really described is man's crowning victory over nature's obstacles to his happiness. Once this triumph is achieved, it follows that man's program for the future is to take advantage of his success and establish on this earth the promised paradise. This, however, is much easier said than done, because—and here we have the real problem—the world built and sustained by technology comprises a tragic paradox. Indeed, if it is true that in such a world nature's obstacles are or may be abolished, it is none the less true that that same world begets, in its turn, new obstacles of another kind, which, because of that, may not be overcome by the same means as the others. Man, consequently, is faced with a whole new set of pressures due to adverse conditions of his own making. And in this respect we must carefully avoid the snare of incriminating technology on that score. That amounts to fleeing from truth by the handy means of shunning the responsibility of whatever befalls.

Technology of itself is neither good nor evil, and to blame it is like reproaching the iceberg for having sunk the *Titanic*. Obviously, the sin is not to be found in technology but in the use to which it may be put. But here again we must be careful, because it is not a question—as it is frequently believed—of merely the reckless carelessness in the production and use of technical devices which has brought about ungovernable cities, pollution, and ecological unbalance. These and other, similar real tremendous evils, notwithstanding the gigantic task involved in curbing and remedying them, do not, however, involve the true problem. After all, they are basically technical problems to be technically solved, though it will undoubtedly mean deep change in social, political, and economic structures.

The great sin—let us call it that—is in having recourse to technology to attain ends alien to its native objective of making man's happiness a historical possibility by freeing him from want. To succumb to the temptations, for example, of using the great power of technological enterprise to achieve predominance at the cost of the well-being of the great masses in underdeveloped areas; to implement heinous doctrines involving the extermination of a whole people; or, more subtly, to transform a world so patiently and fearlessly built for man's well-being into its exact opposite—these are corruptions of man's historic imperative. In these examples we have instances of man's own high treason against his historic struggle to liberate himself from his original, animal condition. The propitious and beneficial artificial, man-made world is in jeopardy of becoming an ugly and brutish world in which man will be but a slave to his victory of freedom over nature, his most glorious historical achievement.

Clearly, such a peril now summons man to embark on a bold new enterprise comparable in design and dimension to the war he waged against nature's original opposition, and, consequently, to find the adequate technology to attain victory. We do not have time now to further dwell on this most vital question and may only indicate that such a technology must be addressed to the conquest of internal nature as the other was addressed against the hazards of external nature. Because if in this way man became the master of the universe, it is only too patent that he has yet to become the master of himself. Let us, then, put an end to our reflections by saying that man's project of life for the future, whatever else it might be, must include a program of self-restraint, education, and love, so that, in conquering what I do not hesitate to call spiritual innocence, he may regain paradise lost. The alternative—it goes without saying—is Big Brother or, perhaps mercifully, thermonuclear apocalypse.

FOOTNOTES

[1] Oswald Spengler, *Der Mensch und die Technik* (1931).
[2] José Ortega y Gasset, "Meditación de la técnica," *supra*, I.
[3] It is hardly necessary to say that when we speak about man's needs we are not limiting them to his organic or biological wants. On the contrary, the gratification of these is but the previous condition for the truly human necessities to appear, necessities completely superfluous from the biological point of view and meaningless to

an animal. On luxury as a chief objective of technology, cf. Ortega y Gasset, op. cit., II; Spengler, op. cit., V. 10.

4 The use of universe and world as synonymous is equivocal. Later (note 12, *infra,*), we shall understand the reason for that usage. The true meaning of world is, however, preserved in common speech when, for instance, we say of a friend that his work and family are "his world," or when, in reference to the tastes and habits of a man in society we talk of him as a "man of the world."

5 It is worthwhile noticing that in this twofold dichotomy we find the origin of that extraordinary sentiment that has accompanied man for centuries, namely, that he is a stranger in the universe and a rebel against its divinely ordered disposition—the sentiment of alienation conceived by religion as that of guilt derived from original sin.

6 Cf. Ortega y Gasset, op. cit., III.

7 Robert Jay Lifton, in his interesting book *Boundaries: Psychological Man in Revolution* (Vintage, New York, 1970), describes contemporary man as "Protean Man" in view of his extreme social mobility. But this seems to me singularly shallow. Man is "protean" not by historical accident but by constitution.

8 This obviously cannot mean that man will no longer *have* necessities. It means that he will no longer be in need, as explained in IV, *supra.*

9 All this explains the conception of God as, precisely, the non-necessitated entity per se, but it also reveals the goal of man's true ambition.

10 Such is the case of the "true life" imagined by certain Eastern cultures where happiness may only be attained in afterlife on condition of dissolving individuality in the Great Universal Whole. The corresponding technology of such a project of life will not attempt to overcome natural obstacles and it will consist in passive acts such as contemplation, inactivity, and self-sacrifice and immolation.

11 First published in Spanish (Fondo de Cultura Económica, México, 1958) and somewhat enlarged—published later in English (Indiana University Press, Bloomington, 1961).

12 That is the reason why the terms universe and world can be and are used as synonymous (cf. note 4, *supra*). In the light of what has just been explained, it will not appear outlandish to say that Columbus' voyage was truly a voyage into space, while contemporary space trips are voyages not out of the world. The truth of this apparent paradox is reflected in the otherwise alarming indifference towards those trips, except as admirable mechanical achievements.

13 It will be clear that the tremendous implication in this mechanical conception of the universe is that man, not God, is its true master, and here is disclosed the real and ultimate allurement in the project of life of Western culture. Little wonder, then, that authentic religious sentiment has always been wary of science as a manifestation of Satan's pride and has linked technology to the diabolical arts.

14 Quite obviously, the whole of our present inquiry implies an underlying conception of universal history, which would be interesting —or so we think—to describe at leisure on another occasion.

TECHNOLOGY AND EVOLUTION

SIR PETER MEDAWAR *is best known as a pioneer in the field of tissue transplants and immunity. For his research revealing the possibility of acquired immunological tolerance in animals, with applicability to human grafting and organ transplants, he was awarded (with Professor Sir MacFarlane Burnet of Melbourne University) the Nobel Prize for Medicine in 1960.*

He was director of Britain's National Institute for Medical Research (1962–71) and is presently engaged in cancer research at The Clinical Research Centre, Harrow, Middlesex.

In addition to his numerous scientific papers, he has published several collections of philosophical essays, including THE UNIQUENESS OF THE INDIVIDUAL (1957), THE FUTURE OF MAN (1960), THE ART OF THE SOLUBLE (1967), INDUCTION AND INTUITION IN SCIENTIFIC THOUGHT (1969), *and* THE HOPE OF PROGRESS (1972)

I

THE USE OF TOOLS has often been regarded as the defining characteristic of *Homo sapiens*, i.e., as a taxonomically distinctive characteristic of the species. But in the light of abundant and increasing evidence that subhuman primates and even lower animals can use tools, the view is now gaining ground that what is characteristic of human beings is not so much the devising of tools as the communication from one human being to another of the know-how to make them. It was not so much the devising of a wheel that was distinctively human, we may suggest, as the communication to others particularly in the succeeding generation, of the know-how to make a wheel. This act of communication, however rudimentary it may have been—even if it only took the form of a rudely explanatory gesture signifying "It's like this, see," accompanied by a rotatory motion of the arm—marks the beginning of technology, or of the science of engineering.

Everyone has observed with more or less wonderment that the tools and instruments devised by human beings undergo an evolution themselves that is strangely analogous to ordinary organic evolution, almost as if these artifacts propagated themselves as animals do. Aircraft began as birdlike objects but evolved into fishlike objects for much the same fluid-dynamic reasons as those which caused fish to evolve into fishlike objects. Bicycles have evolved and so have automobiles. Even toothbrushes have evolved, though not very much. I have never seen Thomas Jefferson's toothbrush, but I don't suppose it was very different from the one we use today; the Duke of Wellington's, which I *have* seen, certainly was not. To some Victorian thinkers, facts like these served simply to confirm them in the belief that evolution was the fundamental and universal modality of change. The assimilation of technological to ordinary organic evolution was not wholly without substance, because all instruments that serve us are functionally parts of ourselves. Some instruments, like spectrophotometers, microscopes, and radiotelescopes, are sensory accessories inasmuch as they enormously increase sensibility and the range and quality of the sensory input. Other instruments, like cutlery, hammers, guns, and automobiles, are accessories to our effector organs—not sensory but motor accessories.

A property that all these instruments have in common is that they make no functional sense except as external organs of our own: all sensory instruments report back at some stage or by some route through our ordinary senses. All motor instruments receive their instructions from ourselves.

It was for reasons like this that the great actuary and demographer Alfred J. Lotka invented the word "exosomatic," to refer to those instruments which, though not parts of the body, are nevertheless functionally integrated into ourselves. Everybody will have realized from personal experience how closely we are integrated psychologically with the instruments that serve us. When a car bumps into an obstacle, we wince more from an actual referral of pain than from a sudden premonition of the sour and skeptical face of an insurance assessor. When the car is running badly and labors up hills, we ourselves feel rather poorly, but we feel good when the car runs smoothly. Wilfred Trotter, the surgeon, said that when a surgeon uses an instrument like a probe he actually refers the sense of touch to its tip. The probe has become an extension of his finger.

I do not think I need labor the point that this proxy evolution of human beings through exosomatic instruments has contributed more to our biological success than the conventional evolution of our own, or endosomatic, organs. But I do think it is worthwhile calling attention to some of the more striking differences between the two.

II

By far the most important difference is that the instructions for making endosomatic parts of ourselves, like kidneys and hearts and lungs, are genetically programmed. Instructions for making exosomatic organs are transmitted through non-genetic channels. In human beings, exogenetic heredity—the transfer of information through non-genetic channels—has become more important for our biological success than anything programmed in DNA. Through the direct action of the environment, we do in a sense "learn" to develop a skin thicker on the soles of our feet than elsewhere. But information of this kind cannot be passed on genetically, and there is indeed no known mechanism by which it could be. It is only in exosomatic heredity that this kind of transfer can come about. We can learn to make and wear shoes and pass on this knowledge ready made to the next generation. Indeed, we even pass on the shoes themselves.

There is no learning process in ordinary genetic heredity: we can't teach DNA anything, and there is no known process by which the translation of the instructions it embodies can be reversed. No information that the organism receives in its lifetime can be imprinted upon the DNA, but in exogenetic heredity we can and do learn things in the course of life which are transmitted to the succeeding generation; thus exogenetic heredity is Lamarckian or instructional in style, rather than Darwinian, or selective. By no manner of means can the blacksmith transmit his brawny arms to his children, but there is nothing to stop his teaching his children his trade so that they grow up to be as strong and skillful as himself.

The evolution of this learning process and the system of heredity that goes with it represents a fundamentally new biological stratagem—more important than any that preceded it—and totally unlike any other transaction of the organism with its environment. In ordinary, endosomatic evolution and in cognate processes such as the so-called "training" of bacteria and, in immunology, antibody formation, we are dealing with what are essentially *selective*, as opposed to instructive, phenomena. The variants that are proffered for selection arise either by some random process such as mutation or by a process which it is not paradoxical to describe as a "programmed" randomness. By a "programmed randomness" I mean a state of affairs in which the generation of diversity is itself genetically provided for. Mendelian heredity provides for the preservation of genetic diversity for an unlimited period.

Another important difference is this. Genetic evolution is conceivably reversible, just as it is thermodynamically conceivable that a kettle put on a lump of ice will boil. It's very unlikely, that's all. On the other hand, exosomatic evolution is quite easily reversible: everything that has been achieved by it can be lost or not reacquired. This is what specially frightens us when we contemplate the consequences of some particularly infamous tyranny that threatens to interrupt the cultural nexus between one generation and the next. This reversion to a cultural Stone Age is what each political party warns us will be the inevitable consequence of voting for the other. To bring the idea of reversibility to life, one should contemplate the plight of the human race if for any reason it did have to start again from scratch on a desert island: it is not heaven, but the old Stone Age, that lies about us in our infancy.

III

I have been looking around in my mind for some one word or phrase to epitomize what I understand by our human inheritance through non-genetic channels—through indoctrination, that is, and the conscious transfer of information by word of mouth and through books. Karl Popper's[1] new book *Objective Knowledge* supplied the answer ready made. Let me therefore introduce you to Popper's concept of a "third world."

According to the philosophic views we specially associate with the name of George Berkeley, the apparently "real" world about us exists only through and by virtue of our apprehension of it. Thus sensible things and material objects generally exist only as representations or conceptions or as "ideas" in the mind—hence the name "idealism." Berkeley argued persuasively, but Boswell very well knew that Berkeley's argument was of just the kind that would enrage Dr. Johnson. When Boswell teasingly said it was impossible to refute Berkeley's beliefs, "I refute it *thus*," said Johnson, kicking a large stone so violently that he "rebounded" from it, thus simultaneously refuting Berkeley and confirming Newton's third law of motion (the one about actions' having equal and opposite reactions).

However, even those who take a sturdily Johnsonian view of Berkeley's philosophy as it relates to the real world of material objects sometimes hold a Berkeleyan, or subjectivist, view of things of the mind. They tend to believe that thoughts exist by reason of being thought about, conceptions by virtue of being conceived, and theorems because they are the products of deductive reasoning.

Popper's new ontology[2] does away with subjectivism in the world of the mind. Human beings, he says, inhabit or interact with three quite distinct worlds: World 1 is the ordinary physical world, or world of physical states; World 2 is the mental world, or world of mental states; the "third world" (one can see why he now prefers to call it World 3) is the world of actual or possible objects of thought—the world of concepts, ideas, theories, theorems, arguments, and explanations—the world, let us say, of all artifacts of the mind. The elements of this world interact with each other much like the ordinary objects of the material world: two theories interact and lead to the formulation of a third; Wagner's music influenced Strauss's and his in turn all music written since. Again, I mention for what it may be worth that we speak of things of the mind in a revealingly objective way: we "see" an argument, "grasp" an idea, and "handle" numbers, expertly or inexpertly as the case may be. The existence of World 3, inseparably bound up with human language, is the most distinctively human of all our possessions. This third world is not a fiction, Popper insists, but exists "in reality." It is a product of the human mind but yet in large measure autonomous.

This was the conception I had been looking for: the third world is the greater and more important part of human inheritance. Its handing on from generation to generation is what above all else distinguishes man from beast.

Popper has argued strongly that, although the third world is a human artifact, yet it has an independent objective existence of its own—and is indeed quite largely autonomous. I have already pointed out that the third world undergoes the kind of slow, secular change that is described as evolutionary,[3] i.e., is gradual, directional, and integrative in the sense that it builds anew upon whatever level may have been achieved beforehand. The continuity of the third world depends upon a non-genetical means of communication and the evolutionary change is generally Lamarckian in character, but there are certain obvious parallels between exosomatic evolution and ordinary, organic evolution in the Darwinian mode. Consider, for example, the evolution of aircraft and of automobiles. A new design is exposed to pretty heavy selection pressures through consumer preferences, "market forces," and the exigencies of function, by which I mean that the aircraft must stay aloft and the cars must go where they are directed. A successful new design sweeps through the entire population of aircraft and automobiles and becomes a prevailing type, much as jet aircraft have replaced aircraft propelled by airscrews.

I hope it is not necessary to say that the secular changes undergone by the third world do not exemplify and are not the product of the workings of great, impersonal historical or sociological forces. Just as the third world objectively speaking is a human artifact, so also are all the laws and regulations which govern its transformations. The idea that human beings are powerless in the grip of vast historical forces is in the very deepest sense of the word nonsensical. Fatalism is the most abject form of the aberration of thought which Popper calls "historicism." Its acceptance or rejection has not depended upon cool philosophic thought but rather upon matters of mood and of prevailing literary fashion. There was quite a fashion for fatalism in late Victorian and Edwardian England, admirably exemplified by Omar Fitzgerald's famous stanza:

> 'Tis all a Checkerboard of Nights and Days
> Where Destiny with Men for Pieces plays
> Hither and thither moves, and mates and slays
> And one by one back in the Closet lays.

This is a comfortable doctrine, in so far as it spares us any exertion of thinking, but we may well wonder why it was so prevalent in late Victorian and Edwardian England. The answer surely is that it fits very well with that high-Tory and latterly Fascist philosophy according to which, regardless of his upbringing, a man's breeding and genetic provenance fix absolutely his capabilities, his destiny and his deserts: a man not born a gentleman or, e.g., a German, could only at best merely simulate gentility or Germanity.

This kind of fatalism sounds very dated today, but we should ask ourselves very seriously whether there is not a tendency today to take the almost equally discreditable view that the environment has already deteriorated beyond anything we can do to remedy it—that man has now to be punished for his abandonment of that Nature which, according to the scenario of a popular Arcadian daydream, should provide for all our reasonable requirements and find a remedy for all our misfortunes. It is this daydream that lies at the root of today's rancorous criticism of science and the technologies[4] by people who believe, and seem almost to hope, that our environment is deteriorating to a level below which it cannot readily support human life. My own view is that these fears are greatly and unreasonably exaggerated.[5] Our present dilemma has something in common with those logical paradoxes that have played such an important part in mathematical logic. Science and technology are responsible for our present predicament but offer the only means of escaping the misfortunes for which they are responsible.

The coming of technology and the new style of human evolution it made possible was an epoch in biological history as important as the evolution of man himself. We are now on the verge of a third episode, as important as either of these: that in which the whole human ambience—the human house—is of our own making and becomes as we intend it should be: a product of human thought—of deep and anxious thought, let us hope, and of forethought rather than afterthought. Such a union of the first and third worlds of Popper's ontology is entirely within our capabilities, provided it is henceforward made a focal point of creative thought.

The word "ecology" has its root in the Greek word "oikos," meaning "house" or "home." Our future success depends upon the recognition that household management in this wider sense is the most backward branch of technology and therefore the one most urgently in need of development. An entirely new technology is required which is founded upon ecology in much the same way as medicine is founded on physiology. A blueprint for such a technology is described in the book *Only One Earth*, by Barbara Ward and René Dubos,[6] written in preparation for the United Nations World Conference on the Human Environment, held in Stockholm last year. If this new technology comes into being, I shall be completely confident of our ability to put and keep our house in order.

FOOTNOTES

1 Sir Karl Popper, the great philosopher and author of *The Open Society and Its Enemies.*
2 Karl R. Popper. *Objective Knowledge—An Evolutionary Approach* (Clarendon Press, Oxford, 1972).
3 P. B. Medawar. *The Future of Man* (Methuen, London; Basic Books, New York; 1959).
4 P. B. Medawar and J. S. Medawar, in *Civilisation and Science; in Conflict or Collaboration?* Ciba Foundation Symposium (North-Holland, London and New York, 1972).
5 P. B. Medawar. *The Hope of Progress* (Methuen, London, 1972; Doubleday, New York, 1974).
6 Barbara Ward and René Dubos. *Only One Earth: The Care and Maintenance of a Small Planet* (Norton, New York, 1972).

TECHNOLOGY AND THE
LIMITS OF KNOWLEDGE

ARTHUR C. CLARKE, *one of the best-known writers of science fiction and one of the most imaginative and best-informed explorers of the future, has written some forty-five books, more than thirty of them on space travel and space science. His first science fiction,* PRELUDE TO SPACE *(1951), was preceded by two non-fiction books on space travel: an introduction to astronautics,* INTERPLANETARY FLIGHT *(1950), and* THE EXPLORATION OF SPACE *(1951). His most recently published works are* THE LOST WORLDS OF 2001 *(1971), a short-story collection titled* A WIND FROM THE SUN *(1972), and* RENDEZVOUS WITH RAMA *(1973).*

*Clarke collaborated with film producer Stanley Kubrick to make one of Clarke's short stories, "*THE SENTINEL,*" into the movie* 2001: A SPACE ODYSSEY.

In 1945, Clarke proposed the concept of a communications satellite to relay global radio and television signals. The idea was dismissed at the time, but in 1965 the first such satellite was launched by COMSAT, actually using the orbit that had been plotted by Clarke.

THE TWIN SUBJECTS of this talk are *technology* and *knowledge*, and whenever I hear that second word I am reminded of a little poem popular at Oxford about a hundred years ago:

> I am the Master of this College;
> What *I* don't know isn't knowledge.

This claim was, of course, immediately challenged by a rival establishment:

> In all Infinity
> There is no-one so wise
> As the Master of Trinity.

Unless my memory is betraying me yet again, the modest first couplet emanated from Balliol and was attached to Benjamin Jowett, the theologian and Greek scholar.

Today, of course, a man like Dr. Jowett lies squarely on the far side of the famous Culture Gap. Most of today's knowledge consists of things that he didn't know, and couldn't possibly have known. This is not because of the sheer increase in knowledge, though that has been enormous. But the very center of gravity of scholarship has now moved so far that there are vast areas where any high school dropout is better informed than the most highly educated man of a hundred years ago.

Much of this change may be linked with the other gentleman I mentioned just now: the Master of Trinity, than whom, et cetera. The most famous holder of this post was J. J. Thomson, discoverer of the electron; and *that* discovery marks the great divide between our age and all ages that have gone before. It transformed technology, and it transformed knowledge.

The electronic revolution, and the devices it has spawned, is now changing the face of our world and will determine the very structure of future society. And the discovery of the electron led, of course, directly to modern physics and the picture of the universe we have today—so much more complex and fantastic than could possibly have been imagined by any philosopher of the past.

One is almost tempted to argue that most *real* knowledge is a by-product of technology, but of course this is an exaggeration. Much that we know about the world around us has been derived over the centuries by simple naked-eye observation: in some important fields, like botany and zoology, this is still partly true. Yet, even here, we could never have understood the facts of simple observation without the technology represented by the microscope and the chemical laboratory. It can be argued that we do not really know anything until we understand it; mere description is not enough. Ancient naturalists such as Aristotle and Pliny recorded many of the basic facts of genetics; it is only in our time that the secret of the DNA molecule was uncovered, after a gigantic research effort involving every weapon in the technological armory from computers to electron microscopes.

There are those who think that this is a pity, and who somehow feel that knowledge is "purer" in direct proportion to its lack of contamination with technology. This—literally!—mandarin attitude is a consequence, as J. D. Bernal has remarked, of "the breach between aristocratic theory and plebeian practice which had been opened with the beginning of class society in early civilization and which had limited the great intellectual capacity of the Greeks."[1]

This failure of the Greeks—and the Chinese—to fuse technology and knowledge in a truly creative manner is one of the great tragedies of human history; it lost us at least a thousand years. Both these great civilizations had plenty of technology, some of a very high order, as Joseph Needham and others have shown. Nor did the Greeks despise it, as is often imagined; the myth of Daedalus and the reality of Archimedes show their regard for sophisticated mechanics.

Yet, somehow, these brilliant minds—of whom it has also been truly said that they invented all known forms of government and couldn't make one of them work—missed the breakthrough into experimental science; that had to wait for Galileo, two thousand years later. How near the Greeks came to the modern age you can see, if you have sufficient influence and persistence—it took me three visits and a letter from an admiral—in the basement of the National Museum of Athens. For there, tucked away in a cigar box, is one of the most astonishing archaeological discoveries of all time, the fragments of the astronomical computer found by sponge divers off the island of Antikythera in 1901. To quote from Dr. Derek Price: "Consisting of a box with dials on the outside and a very complex assembly of gear wheels mounted within, it must have resembled a well-made *18th Century* (my italics) clock. . . . At least twenty gear wheels of the mechanism have been preserved, including a very sophisticated assembly of gears that were mounted eccentrically on a turntable and probably functioned as a sort of epicyclic or differential gear system."[2]

Looking at this extraordinary relic is a most disturbing experience. Few activities are more futile than the "What if . . ." type of speculation, yet the Antikythera mechanism positively compels such thinking. Though it is over two thousand years old, it represents a level which our technology did not reach until the eighteenth century.

Unfortunately, this complex device merely described the planets' apparent movements; it did not help to *explain* them. With the far simpler tools of inclined planes, swinging pendulums, and falling weights, Galileo pointed the way to that understanding, and to the modern world.

If the insight of the Greeks had matched their ingenuity, the industrial revolution might have begun a thousand years before Columbus. By this time we would not merely be pottering around on the moon; we would have reached the nearer stars.

One of the factors which has caused this gross mismatch between ability and achievement is what might be called intellectual cowardice. In the extreme case, this is best summed up by that beloved cliché from the old-time monster movies: "This knowledge was not meant for man." Cut to the horrified faces of the villagers, as the mad scientist's laboratory goes up in flames. . . .

The non-celluloid version is a little less dramatic. It consists of assertions that something can never be known, or done, rather than that it *shouldn't* be. But often, I think, the underlying impulse is fear, even if the only danger is the demolition of a beloved theory. Let me give some examples which are relevant to the theme of this talk.

It's grossly unfair to judge anyone by a single piece of folly; few of us would survive such a critique. But I have never taken Hegel seriously—and have thus saved myself a great deal of trouble—because of the *Dissertation on the Orbits of the Planets*, which he published in 1801.

In this essay, he attacked the project then under way to discover a new planet occupying the curious gap between

Mars and Jupiter. It was philosophically impossible, he explained, for such a planet to exist. . . . By a delightful irony of fate, the first of the asteroids had *already* been discovered a few months before Hegel's unfortunate essay appeared. I do not know if he issued a revised edition, but Gauss remarked sarcastically that this paper, though insanity, was pure wisdom compared to those that Hegel wrote later. . . .[3]

Some of my best friends are Germans, but I cannot resist quoting an even more splendid specimen of Teutonic myopia. When Daguerre announced his photographic process in 1839, it created such a sensation that some people simply refused to believe it. A Leipzig paper "found that Daguerre's claims affronted both German science and God, in that order: 'The wish to capture evanescent reflections is not only impossible, as has been shown by thorough German investigation, but . . . the will to do so is blasphemy. God created man in his own image, and no man-made machine may fix the image of God. . . . One can straightway call the Frenchman Daguerre, who boasts of such unheard-of things, the fool of fools.'"[4]

One would like to know more about the subsequent career of this critic, whom I have resurrected not merely for comic relief but because he provides an excellent introduction to a much more instructive *débâcle*. This time, I am happy to say, the culprit is a Frenchman; I would not like anyone to accuse me of nationalistic bias. He is the philosopher Auguste Comte.

In the second book of his *Course of Positive Philosophy* (1835) Comte defined once and for all the limits of astronomical knowledge. This is what he said about the heavenly bodies:

We see how we may determine their forms, their distances, their bulk, their motions, but we can never know anything of their chemical or mineralogical structure; and much less, that of organized beings living on their surface. We must keep carefully apart the idea of the solar system and that of the universe, and be always assured that our only true interest is with the former. . . . The stars serve us scientifically only as providing positions. . . .

Elsewhere, Comte pointed out another "obvious" impossibility; we could never discover the temperatures of the heavenly bodies. He thus ruled out even the *theoretical* existence of a science of astrophysics; it is therefore doubly ironic that within half a century of his death, most of astronomy was astrophysics, and scarcely anyone was concerned with the solar system, which he claimed was "our only true interest."

The demolition of Monsieur Comte was produced by a single technological development: spectroscopy. I don't see how we can blame Comte for not imagining the spectroscope; who could possibly have dreamed that a glass prism would have revealed the chemical composition, temperature, magnetic characteristics, and much else, of the most distant stars? And can we be sure that, even now, we have discovered all the ways of extracting information from a beam of light?

Let us think about light for a moment, as the development of optics provides the most perfect example of the way in which technology can expand the frontiers of knowledge. Vision is our only long-range sense—unless one accepts ESP—and until this century everything that we knew about the universe was brought to us on waves of light. If you doubt this, just close your eyes—or reread H. G. Wells's most famous short story:

Núñez found himself trying to explain the great world out of which he had fallen, and the sky and mountains and sight and such-like marvels, to these elders who sat in darkness in the Country of the Blind. And they would believe and understand nothing whatever he told them. . . . For fourteen generations these people had been blind and cut off from all the seeing world; the names of all things of sight had faded and changed . . . and they had ceased to concern themselves with anything beyond the rocky slopes above their circling wall. Blind men of genius had arisen among them and questioned the shreds of belief and tradition they had brought with them from their seeing days, and had dismissed all these things as idle fancies, and replaced them with newer and saner explanations. Much of their imagination had shrivelled with their eyes. . . .

With the genius of the poet he pretended he wasn't, Wells created in this story a universal myth. ". . . *their imagination had shrivelled with their eyes.*" And, on the contrary, how ours has expanded, not only with our eyes, but even more with the instruments we have applied to them!

Galileo and the telescope is the classic example; who has not envied him, for his first glimpses of the mountains of the moon, the satellites of Jupiter, the phases of Venus, the banked starclouds of the Milky Way? During those few months in 1609–10 there occurred the greatest expansion of man's mental horizons that has ever occurred in the whole history of science. The tiny, closed cosmos of the medieval world lay in ruins, its crystalline spheres shattered like the fragments of some discarded nursery toy. Which, in a sense, it was, being one of the childish things our species had to put aside before it could face reality. There will be other sacrifices to come.

One of the most remarkable things about the technology of handling light is the simplicity of the means involved, compared with the far-reaching consequences. To give a humble but revolutionary example, consider spectacles.

The scholars of the ancient world, struggling to read by oil lamps and candles, must often have been functionally blind by middle age—especially as the manuscripts they studied were usually designed for art rather than legibility. The invention of eyeglasses (circa 1350) may well have doubled the intellectual capacity of the human race, for with their aid a man need no longer give up his work just when he was entering his most productive years. I don't know if this is an original idea—or whether it has already been refuted—but one could make a case for spectacles being a prime cause of the Renaissance.[5]

It is hard to think of a simpler piece of apparatus than a lens; yet what wonders it can reveal! Few people realize that the remarkable Dutch observer Van Leeuwenhoek discovered bacteria *with an instrument consisting of a single lens!* His "microscopes" were nothing more than beads of glass mounted in metal plates, yet they opened up a whole universe. Unfortunately, Van Leeuwenhoek was such a genius that no one else was ever able to match his skill, and the microscope remained little more than a toy; for its full use, it had to wait until Pasteur, two hundred years later.

And here is another of the great ifs of technology. Suppose Van Leeuwenhoek's observations had been followed up; then the germ theory of disease—often suggested but never proved—might have been established in the seventeenth century instead of the nineteenth. Hundreds of millions of lives would have been saved—and by this time the population explosion would have come and gone. Human civilization would by now have collapsed—or it would have safely passed through the crisis which still lies ahead of us—and for which, if you will excuse me, I have coined the word *Apop*aclypse.

The microscope and the telescope, both born about the

same time, thus have sharply contrasting histories. The microscope remained a toy—the plaything of rich (well, fairly rich) amateurs like Samuel Pepys.[6] But the telescope, from the moment it was introduced, started a revolution in astronomy that has continued to this day.

Many years ago, however, the telescope came up against an apparently fundamental limit—that of *practical* magnifying power. Because of the wave nature of light, there is no point in using very high magnifications; the image simply breaks up, like an overenlarged newspaper block. Still, this natural limit is a very generous one. In theory, the Mount Palomar 200-inch would permit an incredible 20,000 power, which would bring the moon to within ten miles.

Alas, this delightful fantasy is frustrated by the medium through which the light must pass—the few dozen miles above the observatory. A star image can travel intact for a million million million miles, only to be hopelessly scrambled during the last microseconds of its journey by turbulence in the earth's atmosphere.[7] To the optical astronomer, all too often, the medium is the mess.

Even under the rare conditions of virtually perfect seeing at mountaintop observatories, the highest magnification that can ever be used is only about 1,000. This means that under favorable conditions the smallest object that can be seen on the moon is about half a mile across, and on Mars about fifty miles across. But these figures are very misleading, because contrast plays a vital role. Lunar contrasts can be very high, owing to the starkness of the shadows. Mars contrasts are very low, making its surface features difficult to see and still harder to draw and photograph.

This tantalizing state of affairs led to one of the most famous, entertaining, and perhaps tragic episodes in the history of astronomy; I refer to the long controversy over the Martian canals, which has been finally settled only during the past few months. It is an example of what can happen when the desire for knowledge outruns the technology of the time.

Though he was not the first man to "observe" the canals, Percival Lowell was certainly the man who put them on the map—and I use that phrase with malice aforethought. Carl Sagan has, perhaps unkindly, referred to Lowell as "one of the worst draftsmen who ever sat down at the telescope"; I have preferred to call him "the man with the tessellated eyeballs."[8]

Whatever Lowell's deficiencies as an observer, there can be no doubt of his ability as a propagandist. In a series of persuasive books, from 1895 onward, he almost singlehandedly laid the foundations of a myth which was gleefully elaborated upon by several generations of science-fiction writers—of whom the most celebrated were Wells, Burroughs, and Bradbury. The ancient sea beds, the vast irrigation system which still brought life to a dying planet, the ruins of cities that would make Troy seem a creation of yesterday—it was a beautiful dream while it lasted, which was until July 15, 1965.

On that day, the overstrained technology of the telescope was surpassed—though by no means superseded—by that of the TV-carrying space probe. Mariner 4 gave us our first glimpse of the real Mars, though by another delightful irony of fate those initial pictures were almost as misleading as Lowell's fantasies. Not until Mariner 9's superb mapping of the entire planet, in 1972, did Mars slowly begin to emerge from the mists as a unique geological entity—and one of our main orders of business in the next hundred years.

In the other direction, down toward the atom, we have also broken through one apparently insuperable barrier after another. The optical telescope and the optical microscope reached their limits at about the same time, since these are both set by the wave nature of light. The wholly unexpected invention of the electron microscope suddenly increased magnifying power a thousandfold, allowing us to view structures of molecular size and producing advances in the understanding of living matter that could have been obtained in no other way.

In the past few years, there has been another breakthrough—I hate having to use this exhausted word, but there are times when there is no alternative. The scanning electron microscope has done something quite new, and wholly beyond the power of the older optical and electronic instruments. By showing minute three-dimensional objects in sharp focus, it has allowed us for the first time to enter—emotionally, at least—the submicroscopic world. When you look at a good S.E.M. photo of some creature barely visible to the eye, you can easily believe that it is really as large as a dog—or even an elephant. There is no sense of scale; it is as if Alice in Wonderland's fantasy has come true and we are able to shrink ourselves down to insect size and have an eyeball-to-eyeball confrontation with a beetle.

The power of technology to change one's intellectual viewpoint is one of its greatest contributions not merely to knowledge but to something even more important: *understanding*. I cannot think of a better proof of this than some remarks made by Apollo 8's William Anders at the signing of the Intelsat Agreement here in Washington on 20 August 1971:

Truly, the most amazing part of the flight was not the moon, but the view we had of the earth itself. We looked back from 240,000 miles to see a very small, round, beautiful, fragile-looking little ball floating in an immense black void of space. It was the only color in the universe—very fragile—very delicate indeed. Since this was Christmas time, it reminded me of a Christmas-tree ornament—colorful and fragile. Something that we needed to learn to handle with care. . . .

Now, the telescope, the microscope, even the rocket have given us only a change of scale or of viewpoint in *space*. In a more modest way, men have been achieving this ever since they started exploring the earth. What is very new in human history is the power to change our outlook on *time*.

The camera was the first breakthrough in this difficult area. From the beginning, the photographic plate could capture a moment out of time, in a manner never even conceived before—witness the quotation I have already given from that Leipzig newspaper. We all know the extraordinary emotional impact of old photographs; this is because they can provide, in a way not possible even to the greatest art, a window into the past. We can look into the eyes of Lincoln or Darwin; but not of Washington or Newton. At least, not yet. . . .

Because the first photographic emulsions were very slow, the earliest glimpses of the past were somewhat extended ones; they lasted minutes at a time. But about a hundred years ago the camera acquired sufficient speed to provide mankind with another wonderful tool; call it an image-freezer, or time-slicer.

It is in fact almost exactly a century ago (1872) that the flamboyant Eadweard (sic) Muybridge solved a famous problem that had baffled every artist since the creators of the first cave paintings. Does a running horse have all four feet off the ground at the same time? Muybridge found that the answer was "yes"; he also discovered that the characteristic "rocking-horse" position shown in innumerable paintings of charging cavalry and Derby winners was nonsense. This caused great heartburning in artistic circles.

Since then, of course, the camera has speeded up many millionfold. Until recently, the ultimate was a photograph which Dr. Harold Edgerton has hanging up in his office at M.I.T. It shows a steel tower surmounted by a globular cloud with three cables leading into it. The ends of the cables are a little fuzzy, which is not surprising. The cloud is an A-bomb, a few microseconds after zero. . . .

Yet now we have far surpassed that. Using laser techniques, a slug of light less than a centimeter long has been stopped in its tracks. I suppose we'll reach the end of the line when someone catches a single photon in mid-vibration. . . .

We are obviously a long way from the tempo of the running horse—just beyond the limits of human perception —or even of the hummingbird's wingbeat, which is still something that the mind can comprehend even if the eye cannot grasp it. Slicing time into thinner and thinner wafers has now led us into the weird world of nuclear phenomena, and perhaps even down to the atomic or granular structure to time itself. There may be "chronons," just as there are photons.

Now that every amateur photographer has a flash gun plugged into his camera, the power to freeze movement no longer seems such a miracle. And, of course, stopping time is not a *wholly* new experience to men, though in the past it was an uncontrollable one. A thunderstorm on a dark night was pre-twentieth-century man's equivalent of a modern strobe-light show, and must have impressed him even if he did not particularly enjoy it.

Worlds may exist which have natural strobe lighting, though until a few years ago not even the most irresponsible of science-fiction writers would have dared to imagine such a thing. For who could have dreamed of a star which switched itself on and off thirty times a second?

The Crab pulsar does just this, and it's strange to think that its flashes might have been discovered years ago—if anyone had been insane enough to look for them with suitable equipment. What would have happened to astronomy, I wonder, if that had been done? Many people would have been convinced that such a flickering star was artificial—and even now, I don't think we should dismiss this explanation. The pulsars may yet turn out to be beacons, and those who protest that they are a very inefficient way of broadcasting have been neatly answered by Dr. Frank Drake. How do we know, he has asked, that there aren't some *stupid* supercivilizations around?

Granted the improbability that it could have survived the initial supernova explosion, it's fascinating to speculate about conditions on a world circling the Crab pulsar. To our eyes, daylight would appear to be continuous, but it would really be 30 cycles per second A.C. A rapidly moving object would break up into discrete images. Something that appeared to be stationary might be really spinning at high speed.

What would be the effect of this on evolution? Could predators take advantage of these weird conditions to deceive their victims? One day I may work this idea up into a story. Meanwhile if Isaac Asimov (to take a name at random) uses it first, you'll know who he stole it from.

So far I have talked about slowing down time, but what about speeding it up? Of course, that's much easier; it requires very simple technology, but lots of patience. Though the results are often fascinating, I do not know if they have yet contributed much to scientific knowledge. Time-lapse films of clouds and growing plants are the best-known examples in this field. Anyone who has watched the fight to the death between two vines, striking at each other like serpents, will have a new insight on the botanical kingdom. And I hope that the meteorologists will learn a great deal from the global cloud-movement films that have been taken from satellites; eventually these may give us a synoptic view of the seasons and even long-term climatic changes.

This was anticipated by two great writers of the last century; first:

. . . Night followed day like the flapping of a black wing . . . and I saw the sun hopping swiftly across the sky, leaping it every minute, and every minute marking a day. . . . The twinkling succession of darkness and light was excessively painful to the eye. Then, in the intermittent darkness, I saw the moon spinning swiftly through her quarters from new to full, and had a faint glimpse of the circling stars. Presently, as I went on, still gaining velocity, the palpitation of night and day merged into one continuous greyness. . . . the jerking sun became a streak of fire, a brilliant arch, in space; the moon, a fainter, fluctuating band. . . . Presently I noted that the sun belt swayed up and down, from solstice to solstice, in a minute or less, and that consequently my pace was over a year a minute; and minute by minute the white snow flashed across the world, and vanished, and was followed by the bright, brief green of spring.

That, as I am sure you have all recognized, is from Wells's first—and greatest—novel, *The Time Machine*. Unlike his space romances, it is a book that could not have been written before the nineteenth century; only then had the geologists finally shattered the myth of Genesis 4004 B.C. and revealed the immense vistas of time that lie in the past—and may lie ahead. The emotional impact of that discovery on the more sensitive Victorians is preserved in Tennyson's famous lines[9]:

There rolls the deep where grew the tree.
O earth, what changes hast thou seen!

There where the long street roars hath been
The stillness of the central sea.

The hills are shadows, and they flow
    From form to form, and nothing stands;
    They melt like mist, the solid lands,
Like clouds they shape themselves and go.

We now know that this poetic vision is a pretty good description of continental drift, suddenly respectable after languishing for years in the wilderness somewhere to the southeast of Velikovsky. And what established continental drift was a series of breakthroughs in technology, quite as exciting as those involved in the exploration of space. We are accustomed to sending probes to the planets. We have now begun to send probes into the past.

I don't mean this literally, of course; I don't believe in time travel, though I understand that Kurt Gödel has shown that it is theoretically possible under certain peculiar and highly impracticable circumstances, involving the annihilation of most of the universe. The best argument against time travel, as has been frequently pointed out, is the notable absence of time travelers. However, a few years ago, one science-fiction writer pointed out a chillingly logical answer to this. Time travelers, like radio waves, may need a receiver . . . and none has been built yet. As soon as one is invented, we may expect visitors from the future . . . and we had better watch out.

Looking into the past, however, does not involve logical paradoxes, and our time probing has brought back knowledge which a few years ago would have been regarded as forever hidden. I wonder what Auguste Comte would have said if one had asked him the chances of finding the age of a random piece of bone, of locating the North Pole a million years ago, of measuring the temperature of the Jurassic ocean, or the length of the day soon after the birth of the moon? I feel quite certain that he would have said such things are as intrinsically unknowable as the composition of the stars. . . .

Yet such knowledge is now ours, and often through methods which, in principle at least, are surprisingly simple. Everyone is aware of the revolution in archaeology brought about by carbon 14 dating. The still more surprising science of paleothermometry depends on similar principles. If you measure the isotope ratios in the skeletons of marine creatures, you can deduce the temperatures of the seas in which they lived. So we can now go back along the cores taken from the ocean bed, and watch the rise and fall of the thermometer as the ice ages come and go, one after another. There is, surely, something almost magical about this. . . .

To track the wanderings of the earth's magnetic poles ages before compasses were invented or there were men to use them, appears equally magical. Yet once again, the trick seems simple—as soon as it is explained. When molten rock cools, which happens continually during volcanic eruptions, it becomes slightly magnetized in the direction of the prevailing field. The tiny atomic compasses become frozen in line, carrying a message which sensitive magnetometers can decipher.

But what about the length of the day millions of years ago? Surprisingly, this requires the simplest technology of all; merely a microscope, and infinite patience.

Just as the growth of a tree is recorded in successive rings, so it is with certain corals. But some of them show not only *annual* layering but much finer bands of *daily* growth. By studying these, it has been discovered that six hundred million years ago the earth spun much more

134

swiftly on its axis; it had a 21-hour day, and there were 425 days in the year.

These remarkable achievements, and others like them, are allowing us to reconstruct the past like a gigantic jig-saw puzzle. Just how far can the process go? Is there *any* knowledge of the past which is forever beyond recovery?

A favorite science-fiction idea—though I have not seen it around recently—is the machine that can recapture images or sounds from the past. Many will consider that this is not science fiction, but fantasy. They may be right, but let us indulge in a little daydreaming.

There used to be a common superstition to the effect that no sound died completely, and that a sufficiently sensitive amplifier could recapture any words ever spoken by any man who has ever lived. How nice to be able to hear the Gettysburg Address, Will Shakespeare at the Globe, the Sermon on the Mount, the last words of Socrates. . . . But the naïve approach of brute-force amplification is of course nonsense; all that you would get is raw noise. Within a fraction of a second, all normal sounds, expanding away from their source at Mach One, become so dilute that their energy sinks below that of the randomly vibrating air molecules. Perhaps a thunderclap may survive for a minute, and the blast wave of Krakatoa for a few hours—but your words and mine last little longer than the breath that powers them. They are swiftly swallowed up in the chaos of thermal agitation which surrounds us; and when you amplify chaos, the result is merely more of the same commodity.

Nevertheless, there is a slight hope of recapturing sounds from the past—when they have been accidentally frozen by some natural or artificial process. This was pointed out a few years ago by Dr. Richard Woodbridge in a letter to the I.E.E.E. with the intriguing title "Acoustic Recordings from Antiquity."[10]

Dr. Woodbridge first explored the surface of a clay pot with a simple phonograph pickup, and succeeded in detecting the sounds produced by a rather noisy potter's wheel. Then he played loud music to a canvas while it was being painted, and found that short snatches of melody could be identified after the paint had dried. The final step—achieved only after a "long and tedious search"—was to find a spoken word in an oil painting. To quote from Dr. Woodbridge's letter: "The word was 'blue' and was located in a blue paint stroke—as if the artist was talking to himself or to the subject."

This pioneering achievement certainly opens up some fascinating vistas. It is said that Leonardo employed a small orchestra to alleviate the Mona Lisa's boredom during the prolonged sittings. Well, we may be able to check this—if the authorities at the Louvre will allow someone to prowl over the canvas with a crystal pickup. . . .

A few months ago I wrote to Dick Woodbridge to find if there had been any further developments in this field—which, it should be pointed out, requires the very minimum of equipment; merely a pair of earphones, a phonograph pickup, a steady hand, and unlimited patience. But he had nothing new to report and ended his letter with a plaint which all pioneers will echo. "The bottom part of an S curve is a lonely place to be!" True, but there's a lot of room down there to maneuver. . . .

There must be better ways of recapturing sound, but I can't imagine what they are. Still less can I imagine any way of performing a much more difficult feat—recapturing *images* from the past. I would not say that it is impossible in principle; every time we use a telescope, we are, of course, looking backwards in time. But the detailed reconstruction of ancient images—paleoholography?—must depend upon technologies which have not yet been discovered, and it is probably futile to speculate about them at this stage of our ignorance.

It may well be hoped that—whatever the enormous benefits to the historian—such powers never become available. There is a peculiar horror in the idea that, from some point in the future, our descendants may have the ability to watch everything that we ever do.

But, because a thing is appalling, it does not follow that it is impossible, as the H-bomb has amply demonstrated. The nature of time is still a mystery; there may yet be ways of seeing the past. Is that any stranger than observing the center of the sun—which is what we are now doing with telescopes buried a mile underground? Surely neutrino astronomy, involving the detection of ghostly particles which can race at the speed of light through a million million miles of lead without inconvenience—is a greater affront to common sense than a simple idea like observing the past.

I seem to be in grave danger, just when I am running out of time, of starting on an altogether new talk: "Knowledge and the Limits of Technology." Which only proves, of course, how difficult it is to separate the two subjects—or to establish limits to either.

In fact, no such limits may exist, this side of infinity and eternity. Those who fear that this is indeed true have often tried to call a halt to scientific research or industrial development; their voices have never been louder than they are at this moment.

Well, there may be limits to growth, in the sense of physical productivity, though in a properly organized world we would still be nowhere near them. But the expansion of knowledge—of information—is the one type of growth that uses no irreplaceable resources, squanders no energy. In fact, in terms of energy, information provides some almost unbelievable bargains. The National Academy of Sciences' recent report on astronomy gave a statistic that I would not credit from a less reputable source. All the energy collected by our giant radiotelescopes during the three decades that have revolutionized astronomy is "little more than that released by the impact of a few snowflakes on the ground."[11]

In the long run, the gathering and handling of knowledge is the only growth industry—as it should be. And to make the enjoyment of that knowledge possible, technology must play its other great role: lifting the burden of mindless toil, and permitting what Norbert Wiener called "the *human* use of human beings."

We are only really alive when we are *aware*—when we are interacting with the universe at the highest emotional or intellectual level. Scientists and artists do this; so, to the limits of their ability, did primitive hunters, whose lives we are now completely reassessing in the light of new knowledge. Anthropologists have just discovered, to their considerable astonishment, that we lost the twenty-hour week somewhere back in the Neolithic. For in optimum conditions, a few hours of hunting/foraging a day were all that was needed to secure the necessities of life; the rest of the time could be spent sleeping, conversing, chewing the fat (literally)—and, of course, thinking.

But, as we have seen, thinking doesn't get you very far without technology. We can thank what I have christened the Agricultural-Industrial Complex for that. Unfortunately, this sinister organization also invented work and abolished leisure, which we are only now rediscovering, after a rather nasty ten thousand years. Hopefully, we will be able to make a safe transition into the post-industrial age, and then the slogan of all mankind will be—if I may change just two words in Wilde's famous aphorism:

WORK IS THE CURSE OF THE THINKING CLASSES.

It is technology, wisely used, that will give us time to think, and an unlimited supply of subjects to think about. And if it leads to our successors, either the intelligent computers or the "Giant Brains" that Olaf Stapledon described in his masterpiece *Last and First Men*, why should that be regarded as a greater tragedy than the passing of the Neanderthals?

Our technology, in the widest sense of the word, is what has made us human; those who attempt to deny this are denying their own humanity. This currently popular "treason of the intellectuals" is a disease of the affluent countries; the rest of the world cannot afford it.

Not long ago, I was driving through the outskirts of Bombay when I noticed a sadhu (holy man) with just two visible possessions. One was a skimpy loincloth; the other, slung round his neck on a strap, was a transistorized loud-hailer.

There, I told myself, goes a man who does not hesitate to use technology to spread his particular brand of knowledge. He has grasped the one tool he needs, and discarded all else.

And that is the true Wisdom—whether it comes from the East, or from the West.

FOOTNOTES

[1] *Science in History*, Vol. II, (Pelican Edition, London, 1969) pp. 375–76.

[2] "An Ancient Greek Computer," *Scientific American*, June 1959. On the copy he sent to me, Derek Price has written hopefully, "Please find some more." I am afraid that the most advanced underwater artifact I have yet discovered is an early nineteenth-century soda-water bottle.

[3] Willy Ley, *Watchers of the Skies* (Viking, New York, 1963), p. 320.

[4] *Light and Film* (Time/Life Books, New York, 1970), p. 50.

[5] As one of the undoubtedly countless examples of the need for eye-glasses in the ancient world, see *Julius Caesar*, Act V, scene II:

> *Cassius:* Go, Pindarus, get higher on that hill;
> My sight was ever thick; regard Titinius,
> And tell me what thou notest about the field.

And if anyone wants to know what Romans would have looked like wearing horn-rimmed spectacles, Phil Silvers has already obliged in *Something Funny Happened on the Way to the Forum*.

[6] On July 26, 1663, Pepys bought a microscope for the "great price" of five pounds, ten shillings. "A most curious bauble it is," which he used "with great pleasure, but with great difficulty."

Let us also never forget that, as President of the Royal Society, Pepys's name appears immediately below Newton's on the title page of the *Principia*.

[7] Very recently, it has been possible to approach the theoretical limits of magnification by the technique known as "speckle interferometry." The 200-inch Hale telescope has been used at an effective focal length of *half a mile* to photograph the disks of giant stars like Antares.

[8] Both libels will be found in *Mars and the Mind of Man* (Harper & Row, 1973).

[9] *In Memoriam.*

[10] *Proceedings of the I.E.E.E.*, August 1969, pp. 1465–66.

[11] *Astronomy and Astrophysics for the 1970's* (NAS, 1973), p. 77.

SECOND SERIES

CREATIVITY AND COLLABORATION

1973–74

AKIO MORITA

JAMES D. WATSON

HUW WHELDON

MOSHE SAFDIE

CASPAR W. WEINBERGER

CREATIVITY IN MODERN
INDUSTRY

AKIO MORITA *Japanese industrialist Akio Morita is President of Sony Corporation, which he founded with Masaru Ibuka in 1946. Before its first decade, Sony had patented a magnetic tape-recording system and developed a popular pocket-sized all-transistor radio, the first of its kind in Japan. While working to refine radio transistors in the 1950s, Sony Corporation scientist Leo Esaki developed the tunnel diode, an advance that was to earn him a Nobel Prize for Physics in 1973. Sony Corporation has engineered the miniaturization of television sets, and a unique Trinitron color system for television. The Tokyo-based company manufactures other consumer electronics products as well, among them videotape recorders, hi-fi components, and dictating machines.*

Mr. Morita founded Sony Corporation of America in 1960 and presently chairs its Finance Committee. He is also a director of IBM World Trade Corporation.

In this Frank Nelson Doubleday Series of lectures this year, I have been asked to take up the topic of creativity in the field of industry and technology.

The word "creativity" has such a broad range of meaning that the definition of the word itself may come into question when we try to talk about it. Since I am not a native of an English-speaking country, I shall not attempt to go back to a discussion of the definition of this word. However, we all know how vital creativity is to the progress of mankind, and, of course, industry could not exist without it.

I have worked for nearly three decades in industry, in the field of electronics, where innovation is its very life. And in this field I have managed a corporation that has featured the creation of new products as a consistent policy. Therefore, when I use the word "creativity," I hope you will understand that it is limited to creativity in our industry and particularly to that in my own organization.

I do not intend to discuss the theoretical definition or implications of this word beyond this scope, but there are some general assumptions about it which I would first like to mention here.

I

When we think of being creative in industry and technology, we associate this quality with inventions. I believe that creativity is the motive force that produces inventions and innovations in concrete form. The ability to think of something that has not been known before is creativity. This is an ability that is unique to man. No matter how much a computer may be developed to accumulate knowledge, and no matter how great its capacity may be for processing the information, it cannot have the ability to be creative, since the quality of being truly creative involves a leap beyond the results of processing existing information.

The leap, which is necessary in creative thinking or invention, might be described in another way: It is giving birth to a new concept beyond a system of theory already known, or discovering a concept that breaks through the walls surrounding the existing theory. This is true creation.

However, it is extremely rare for this type of truly creative idea to be generated under ordinary conditions. The talent to make such a leap, therefore, may be said to come only from a so-called genius.

The 1973 Nobel Prize for Physics was awarded to three scientists, including Dr. Leo Esaki. Dr. Esaki's work on the tunnel effect in semiconductors, for which he received this award, was conducted in the laboratory of my company, Sony Corporation. At that time, we were exerting efforts to develop transistors for practical use, as the first company in Japan to do so, and our major goal was to create good transistors that could be used for radios.

Various experiments were being conducted for this purpose, and Dr. Esaki, as a member of this research team, was experimenting with a wide range of semiconductor ingredients which had never been tried out before. One combination of ingredients, he discovered, showed a characteristic that had not been known before, and he therefore began to pursue this phenomenon thoroughly. I imagine that when he noticed this unique phenomenon, it sparked the thought that it was not an error in measurement but something that could be explained theoretically. He thought that this newly discovered phenomenon was the result of "a tunneling effect of a PN junction in the forward direction," and he pursued very detailed and careful experiments on it. From these experiments, he developed the theory. This is perhaps an example of how something creative was born as a result of a genius-like inspiration coupled with very careful and precise experimentation.

But, as I said before, it is not frequent that a single scientist creates a major breakthrough in industry these days. It is done by many in teamwork.

II

This present age has seen the very high development of all fields of science, the participation of great numbers of persons in new research, and very deep specialization in many disciplines and subdisciplines. Science advances very rapidly and over a wide variety of fields, but only a small percentage of the inventions are really creative these days. In our field of electronics, also, a revolutionary discovery that brings about a great change of generation in our technology does not appear frequently. However, this does not mean that there is no progress in industry.

The present age, in which we are living, is an age of innovation. We constantly see improvements and changes, which may not all be great inventions, but they appear in every aspect of our lives. These may be said to have been brought about by the exercising of man's creativity in its broadest sense. We see this broader creativity not only in the laboratory but also the design office, in the factory, in marketing plans, and in all other aspects of modern industry. Such a broader sense of creativity is always required in industry. But it is not spontaneous very often. The question is how such creativity can be generated. Two factors suggested themselves as generators of creativity. These two factors are (1) the clear establishment of goals and (2) the provision of outside stimulation.

Although the great inventions or discoveries of a very high level, which I described before, are likely to appear suddenly, without incentives supplied from the outside, it is often possible to motivate technologists to become creative, by providing such outside influences as a specific goal or stimulation. The great goal of sending men to the moon was established, and as analyses were conducted to determine what needed to be accomplished, many specific objectives became clear. As each one of the specific objectives was further analyzed, the potential for new creativity was revealed. It may be said that, in most cases in modern industry, this is the process by which the creative quality of man is brought out.

In the case of outside stimulation, we can see many examples. One form of this is learning of other creative results, which happens very often. In our industry, which operates under the principles of free competition, very often the knowledge that a competitor has come out with a new product is enough to stimulate us to create a different idea, uniquely our own, and produce a new product that serves the same purpose better. Also, in other cases, both a goal and the stimulation of competition or other information operate mutually to bring about new inventions.

III

I would like to illustrate this point with a specific example in my company. After we came out with the first fully transistorized black-and-white television set in the world, we began research on color television, which we felt was the natural follow-up of black-and-white.

Ever since the company was founded, the policy of Sony has been to produce products different from those of other companies. At that time, all manufacturers of color television in the world were using the shadow mask system invented by RCA. After studying all systems conceived up to that time, we saw that the Chromatron system invented by Dr. Lawrence was unique in its theory. We exerted our efforts to make this system industrially feasible. After devoting ourselves to tremendous amounts of research and development, we finally succeeded in putting the system on the market. But we learned that it was not economically feasible as a business.

During this time, color television was rapidly gaining popularity in the United States, while in Japan, also, other manufacturers had developed their technology using the shadow mask system. They were all enjoying a booming business.

However, we did not give up. We were determined to develop our own, unique system, and the more other companies succeeded in sales with the shadow mask system, the more it stimulated the engineers of Sony. They finally succeeded in developing an entirely new system, which we named Trinitron.

Mr. Senri Miyaoka, who was a young scientist working on picture-tube research, conceived of a system of three in-line electron beams produced from a single electron gun. The three beams were converged by a single large-diameter electron lens and a pair of electron prisms. This completely new system of electron optics, which had never existed before, was epoch-making in the concept of color picture tubes. Compared to the conventional three-gun system, it produces a far brighter and clearer picture.

One of the main features of the development team for this work was that the three inventors of the Trinitron participated in all stages from basic research to pilot manufacture. Also, those in the production department were brought into this team, to start at the same time on development and production. Some men worked during the day in the planning of production equipment and then worked in the evening in the development department. Moreover, from the very beginning of this project, top management took direct charge, deciding on priorities for personnel, funds, and other matters, and putting the decisions quickly into action.

Only two years after the first idea for the Trinitron was conceived, Sony's new color television was being mass-produced and marketed. In five years since then, over 4.3 million Trinitron sets were marketed (as of November 1973). Mr. Miyaoka was designated by the Institute of Electrical and Electronics Engineers in the United States to receive the Vladimir K. Zworykin Award at their conference in Chicago in June 1974, for his contributions to the development of this new concept in color television display tubes. Also, the National Academy of Television Arts and Sciences conferred on Sony an Emmy Award for 1973 for the Trinitron. These awards may be said to underscore the recognized excellence of the Trinitron system. The development of the Trinitron is a good example of how creativity is strongly motivated by the establishment of a specific goal and stimulated from the outside. However, this is still not the whole story.

IV

Although a specific goal and outside stimulation may strongly motivate creativity, these are not the only factors that enable a creative idea to appear as a product. Much more is needed to develop the creative idea into a successful product.

We can see around the world many examples where a good idea has been developed and marketed, but disappeared without being widely accepted, because the product was not properly designed and produced for consumer use.

Also, there are many examples in which a technology developed by one company or in one nation does not become a viable business, but, after it is transferred to some other company or some other nation, it becomes a practical industry. An invention or a discovery, no matter how good, will not become an industry until it is made into a product of excellent feasibility. Thus, a newly developed item produced through creativity will not lead to a new industry unless excellent manufacturing technology accompanies the effort. If such a new industry does not appear, it will not set off a chain reaction of other innovations leading to other new uses or other new products. This is extremely important to industry and may be considered as the difference between industry and academic science. Creativity in industry, to be brought to fruition, requires the development of the necessary technology in the production process. Beyond the production process, another very important stage is necessary for creativity in industry to become a successful business.

This stage is marketing. A good product that is successfully produced might still disappear because appropriate measures are not taken in the marketing process. The newer and more innovative a product is, the more likely it is that the public might not appreciate it at the beginning.

Here I would like to refer to my experience. In 1950, our company marketed a tape recorder. It was of course the first item of its kind in Japan. Our company had developed our own tape and all the components used in the recorder. When we first created this new product, we had a dream of great fortunes to come. It was the result of costly research and development and the concentrated efforts of our engineering staff. We therefore expected that as soon as we put it on the market, it would sell right away. Inventors are always like that. They have the illusion that if they create a new thing, money will come rolling in. But the realities were opposite to our dreams. It was a very valuable experience to learn that unless the consumer appreciates the new product, no matter how creative the idea is, it merely ends as a toy or an object of curiosity.

We therefore realized that we had to cultivate in the consumer an appreciation for the product. In other words, in order to get the general public to accept the product, we had to go out ourselves to get them to understand everything from the features of the product to its use and purpose.

The market is something that has to be developed, and it must begin with the customer being made to realize the value of the product. Sometimes a new product is planned by conducting market surveys. I believe that, in the case of an entirely new product, a market must be created, not surveyed. Another way to say this is that a new product is the creator of a market, and a new product, which is the fruit of creativity, cannot survive without the creation of a new market.

V

Also, we learned that market creation is by its nature accompanied by education of the public, and that the foundation of this process is good communication. However, good communication with the public cannot be achieved by paying attention to this stage alone. There is a great distance between the creative new idea and the unexpectant public. All steps between these points must have good communication before the public can be given accurate information.

We must start with the creative idea and make sure that accurate and complete information about it flows to all stages of its development into the final product. At the production stage, it is accurate and complete information from all previous stages that helps the staff to produce a reliable product of high quality. Information from the production stage, as well as other, preceding stages, must also be accurately transmitted to the marketing group.

The marketing group can then provide adequate and accurate information through all marketing and servicing channels to the public. Also responses from the public must be fed back through the various stages to points in the organization where such responses should have influence. Again, information about these responses must be complete and accurate. Through this system of good communication, the new product can be improved, good servicing provided, and full public appreciation for it attained.

This complex system involves a very large number of persons. And this means that good communication must flow among the entire intricate network.

Therefore, industry is in constant need of collaboration. Good collaboration and good communication go hand in hand. One cannot exist without the other. Together, when properly managed, they form the arteries of modern industry, giving vitality to all parts of the organization.

VI

In our type of industry, which is especially competitive through the use of technology, it is also extremely important to bring this creativity to life ahead of our competitors. In other words, the time factor is very essential in industry. Shortening the time required for development is a key challenge to all enterprises in an industry such as ours.

If management in a competitive industry expects to take the lead technologically by hiring many scientists, giving them large amounts of money, and merely waiting, it may not arrive at anything creative that would help the company compete. Instead, an organizer is extremely important who can give a clear goal to the scientists and engineers, distribute the various functions among the personnel for the goal, avoiding wasted effort, oversee the entire project to determine what needs to be created in each area of responsibility, and always make clear the direction the group should go. This function is research management. It is here that research management is seen to be an important requisite of industrial creativity.

This is a major responsibility and function of top management. In our company, the development of the Trinitron was serious enough to determine the very survival of our firm. This was why our chief executive, Mr. Masaru Ibuka, took personal charge as project manager from the very beginning.

This was a very appropriate action for this major goal of developing an entirely new color television system in the shortest period of time possible. In order to make sure that the engineering capacity of the company and the necessary funds for their work could be mobilized in any way best suited to the project, it was important to have the chief executive of the company be directly in charge of the effort. He devoted nearly all of his time to this project. By mobilizing to the fullest all necessary aspects of the company for this purpose, we arrived at very successful results in a very short time.

The ability of research management may at times have even greater significance than creativity, because modern science has become divided into so many specialized fields and is moving ahead so rapidly.

VII

Today we live in a different age from the Renaissance, when a creative genius like Leonardo de Vinci appeared, or the age of invention and discovery, in which many men, from the days of Newton to Edison, were active, and during which one man's creativity elevated the entire level of science and scholarship, or produced innovative changes in the structure of society, or generated a major new industry.

Now there is very little chance of this happening. Not only is this true, but also it is extremely rare that even an enterprise is originated by the inventions produced by a single genius. On the other hand, however, present-day society, with its complex network of information channels, provides cooperation among many disciplines. This leads to innovation that coordinates many specialized fields of knowledge, in a way that no single genius could ever have covered in preceding eras.

Creativity in Modern Industry

Up until the nineteenth century, creativity was exercised by individual geniuses, through voluntary effort, without outside incentives or stimulation, and sometimes they even had to create by pushing away obstructions by society. Today, however, it has become possible to stimulate many persons to be creative by providing them with a goal in a broad sense. Although it has become very unlikely that one person would create a new scholastic discipline or revolutionize an industry, as was done by Newton or Edison, it has now become possible to obtain collaboration from all corners of the globe, which neither Newton nor Edison could ever have obtained.

Also, in place of a goal, the knowledge of a creative result can be imparted, which then stimulates entirely different creative ideas. This also is a marvelous fruit of an information-rich society, from which the scientist can benefit.

In modern industry, therefore, we see that many factors are involved in bringing the fruits of creativity to the public. First, an environment that inspires creativity is necessary. This often consists of a specific goal and the stimulation of new information and competition. Then, development and design are needed to make the creative idea appear as a feasible product. The manufacturing process must be fully developed, so that the product can be efficiently produced with consistently high quality. Moreover, the market for the product must be cultivated and properly served so that the creative idea is accepted by the general public through good communication.

Throughout these stages, good collaboration and communication are needed over a very wide range of countless numbers of persons. Perhaps it can be said that the most important matter in modern industry is to get such countless numbers of persons working together in one direction without wasted effort. To bring into collaboration many persons in this way is the function and responsibility of management.

In modern industry, a strong management force is necessary that follows through with a consistent philosophy from the stage of scientific research to the stage of production and marketing. Among the enterprises around the world, the ones with this capacity are the ones that bring to maximum fruition the ability of man to create. They are the ones that continue to grow.

Thus we see that, in this age of collaboration, modern industry can draw from man's creativity far greater fruits for his well-being than ever before in history. This transcends barriers of nation, race, or creed. In its best form, this flowering of creativity is perhaps mankind's greatest hope.

THE DISSEMINATION OF
UNPUBLISHED INFORMATION

JAMES DEWEY WATSON *Biologist James Dewey Watson, in collaboration with English biophysicist Francis H. C. Crick, achieved one of the most dramatic research breakthroughs in modern science: the mapping of the molecular structure of deoxyribonucleic acid, or DNA, the substance of heredity. The scientists' model of the "double helix," presented in 1953, was the key to our understanding of the transmission of heredity.*

The famous team, along with biophysicist Maurice H. P. Wilkins, who had made the X-ray studies of DNA upon which Watson and Crick based their model, were honored with the 1962 Nobel Prize for Medicine and Physiology. In his book THE DOUBLE HELIX *(1968), Watson recounts the discovery process.*

Watson has been a Professor of Molecular Biology at Harvard University since 1955, and in 1968 became director of the Cold Spring Harbor Laboratory, on Long Island, New York.

As a boy in Chicago, I used to equate good manners with wearing rubbers over your shoes, something your parents got mad about when you forgot, but somehow sissy and not at all connected to the world of science that I wanted to be part of. Then I only worried about the limitations of my brain and took comfort that no Emily Post-type manual was about to restrict my future actions. Eccentricity, if not total unpredictability, I regarded the mark of the free mind, with Leslie Howard's portrayal of Henry Higgins pointing the way toward true success. Conventions remained for minor minds unwilling to take chances, and so why pay lip service to forms of behavior that at best generate more hypocrisy?

In my more than awkward moments, however, I would get scared that later I would get hell from people who mattered, but never to the point of learning when to wear a tie or to avoid talking about Roosevelt with my father's Republican relatives. Accepting nonsense to please your elders would inevitably lead to similar compromises in later life. So I had little to talk about with most neighborhood kids, already brainwashed into respectability, and even in college I could communicate only with apparent oddballs equally uninterested in ordinary behavior.

Happily, my first contact with high-power minds, which came in 1947, soon after I went to Bloomington as a graduate student of the geneticist S. E. Luria, confirmed my adolescent fantasies. The truly bright did not live like our relatives or nearby neighbors and wasted little time worrying about how they looked, or the polish on their cars, or whether their lawns were overrun with crab grass. They had the guts to concentrate on the important, knowing that they must banish the pedantic if they were to come up with some new law of nature. And if they seemed contemptuous of those stuck in the recent past, their intolerance was in good cause: no one benefits from false praise, and only if the truth is honestly faced does the possibility exist for rebirth. I thus had no doubt that I had become part of the highest form of human achievement, light-years away from the uninformed prejudices of the poor or the callous self-satisfaction of the educated rich.

Then everyone important in modern genetics was constantly in touch, in the winter by post or by going to informal meetings, while the long summer periods were for planning joint experiments when climbing in the mountains or lying on the beaches. As long as you thought about the gene while eating and drinking, there were no age barriers to keep you from the famous, and if they talked nonsense you did not abort your career by setting them right; just the opposite, since persistent politeness was easily mistaken for vapidity, and if you didn't speak your mind who would know that you could hustle and pounce upon clues that others had missed. To be sure, there were those whose facts frequently faltered, but never trying to make it was a much deeper vice than occasional self-deception.

Top-flight genetics thus never attracted a homogeneous congregation, and nothing so surely failed as attempts to whip its recalcitrants into line. Inconsiderate actions, though too common to be unexpected, were naturally not welcomed, and any time might see one or two outcasts designated as beyond the pale. But if they had new ideas or experiments to report, they eventually slipped back and the nature of their black deeds faded when new forms of arrogance were more than you could take. The essence of the gene, however, demanded bold ways of thinking, and life was too precious to waste worrying too long about how your colleagues behaved.

Now, some twenty-five years later, I often have the sense that I still belong to that half-mad, but uniquely wonderful, playing field of my youth, where the aim was the truth, not money, and where decency always took precedence over cunning. But now I know the matter to be more complex, with success in generating new ideas usually being more than the simple combination of native intelligence and good measure of luck. All too often, science resembles playing poker for very high stakes, where revealing one's hand prematurely only makes sense when you have nothing to show.

The basic dilemma is that there are never enough good new ideas to go around. If most scientists were content to take joy in the work of others and not be judged by their ability to generate new facts or ideas, there would be no problem. But most of the scientists I know chose their profession because they were fascinated by the knowable unknown, always hoping that they could be part of the conquering body. To be sure, we often take joy in the discoveries of others, but often only in proportion to the extent that we were not close to the same objective. Mountains are seldom known by their second ascent even when the ways to the top took very different routes. Likewise, credit in science almost always goes only to the person who first saw what was up.

This constant preoccupation with priority is not merely a matter of pride. If it were, we should worry that we are constantly abetting infantile behavior that takes away much of the inherent joy of science, giving to it characteristics that dominate the rat race of the business world. But we all know too well that the types of jobs we eventually get are very much dependent upon how much we produce. There is little enthusiasm for those who always come in second. The key words are productivity and originality, and if those are lacking, the prospects are poor of finding jobs and research money that will let you continue to do high-powered science. And correspondingly, if one hits the jackpot he is assured of an interesting position together with piles of money to generate new data.

Now, alas, I can see no chance that this inherently stressful situation will ever change. All signs point toward a future where our inherent wealth remains grossly inadequate for our basic social needs, much less for the total flowering of the intellectual life. Thus, science is never likely to be supported on anywhere near the scale that will satisfy its performers, and only a small fraction will ever have the resources they would need for the full expression of their talents.

Thus, not only the self-respect but also the way of life of a scientist depends upon his relative ranking among his peers. As long as he retains the ambition of his youth, he is almost always hoping that he may soon pull off something smashingly big that will intrigue his colleagues. And, giving him the encouragement that will tempt him back to the lab at night is the fact that the time span of a real discovery can be very short, maybe only a few hours to a week or so. Though many years of patient work might be necessary to prepare for the crucial experiment, the more relevant fact may be the brevity of the moment of revelation. One moment nothing, the next everything, or at least until euphoria has worn off and you become stumped by some new puzzle.

Usually you discover you are not the only person working on a given problem. Only when working with trivia can you be sure that you will be alone. In any given branch of science, there usually exist a series of well-defined objectives, which if achieved should explain vast areas of natural phenomena. Everyone in a given area knows them, frequently ranking them in the same order of importance. Very often, however, the top goal is recognized as insurmountable today, maybe having to wait one to several decades before enough groundwork can be laid for the final attack. If so, lesser targets are picked, usually by some form of calculation which combines your inherent curiosity with your chance to get there first. No good can come following a path already well traversed, and most certainly never pick a problem that you feel has a very good chance of being solved by someone else. At best, you would be considered derivative, and at worst you could be ostracized by your peers for poaching.

Unfortunately, in most cases this seemingly black-and-white advice is easier to state than to follow. Since science deals with the unknown, the steps to the top are never clearly laid out, and more than one approach may be necessary before the right one becomes apparent. And since each may require the mounting of a substantial effort, any one person or lab usually does not have the resources to move ahead on all fronts. So when the choice is made, it may not be easy later to change your mind, even when you have suspicions you are facing a very steep wall. Equipment has been bought, assistants hired for quite specific tasks, and enough progress made to make you suspect that, given enough time, you will make it. But, all too often, you then become aware that someone else has the same objective but is using a new approach that might make more sense than yours.

Dissemination of Unpublished Information

Generally, the first reaction to the prospect of being scooped is a combination of despair and hope that your opponent, "X," will fall dead. You may consider giving up, but this could leave you without any tangible results to show for years of toil. Furthermore, you often don't have other goals which really strike your fancy. So you may just plow ahead, fearing someday you will see "X" grinning down from the top. So it is hard not to think about retooling your effort to try the same approach as your competition. Even though you are behind, by being a little more clever you might overtake him. He, of course, might then become hellishly mad, believing correctly that your success is a direct consequence of his having discovered the right way to proceed. So you can be almost certain of making a long-time enemy, the more so depending on the importance of the final breakthrough. Of course, many insiders might know that your competitor deserves most of the credit, but you know of many cases where the also-rans were soon lost from sight.

Often the morality of such situations is judged by how you first realized that "X" found a better approach to your problem. If you learned it from an article in a scientific journal in which "X" published details of his approach, then most people will feel that problem is up for grabs. But if you learned of his trick by hearing him speak informally at a meeting, then you might figure you should let him exploit his observations until his first publication. But if what you have learned totally nullifies your current experiments, you could be driven batty sitting on your hands until your colleague writes up his data, particularly if you figured that he was delaying publication until he had everything sewn up.

Even worse, there can be up to a year-long gap between when a manuscript is submitted for publication and when it finally is printed. In this interval, the manuscript is frequently sent by the editor to referees to see whether it deserves publication. Usually, the best referees are those who work in similar areas and so exactly those people most likely to be in conflict of interest with the data they read. While there is often the feeling that referees should refrain from taking advantage of their position, and not tell their students or colleagues juicy new facts, only an insensitive fool could let his student go on with an experiment when his "insider" information tells him that it no longer makes sense.

Moreover, most manuscripts, even before acceptance, are duplicated, with copies (preprints) given to students in the respective lab(s) and to selected colleagues throughout the world. So, in most cases, many interested people besides the referees "legitimately" know of key results long before they are published. My feeling thus is that what counts is not when some new result is finally published, but when its discoverer sees fit to write it up in manuscript form. For then, unless he is obsessively secretive, his news soon will be dispersed by the verbal grapevine throughout the interested scientific community. Of course, the cautionary refrain is often heard, don't tell So-and-So because he is a thief, but such warnings seldom work, either because someone has passed the news without the enjoining injunction or because the concept of secrecy quickly loses its force when the originator of the news is no longer directly involved in its spread. On the whole, I find it impossible to keep "secrets" unless they are totally without interest, and so I forget them, or because So-and-So's reputation for piracy just can't be avoided.

Essential to the maintaining of some fabric of decency is the unambiguous attribution of the receipt of unpublished information, for example by the direct statement at the start of a paper that the respective experiments "were done with the knowledge of another person's results." This may not always mollify "X," but failure to do so and to write up derivative work as if it were totally original will bring forth cries of murder. But, all too often, attributions of credit are so backhanded that only "insiders" know whether two identical findings were independent events or whether one reflected work only started upon the receipt of a hot rumor. In particular, many late-comers feverishly rush their claims into quick publication so that they can have the same calendar year as the original observation. While, at first, most interested spectators know the copy, with time such unwritten gossip fades and in later years equal credit is frequently extended.

Unhappily, the apparent quality of the science does not always mark the person who saw the truth first. It is not only second-rate scientists who behave improperly, but often those of first rank. Major success early in life all too easily creates the appetite for more of the same, and many of our best scientists compulsively regard all relevant new discoveries as natural objects for exploitation by their own research groups. They know all too clearly that well-financed research empires depend on the constant manufacture of important new data. While there exist many scientific giants who need no help in generating one important new fact after another, we must anticipate that the prospect of loss of face, if not of empire, will subconciously lead many generally honorable men to actions they will later try to forget.

So only the uninformed or the forever naïve expect science to be a totally open discipline whose participants always freely discuss their newest findings. While I know many scientists who instantly blurt out all they know, despising the duplicity that can go with keeping key information back, they may not be in the majority. All of us know enough horror stories to realize that some scientists are best avoided. Even if you don't care whether they try to move in on your own baby, there are times when you must defend your students from their predatory claws. So I have friends who urge their students to lie low when a pirate is on the horizon, knowing all too well that their youthful enthusiasm may become masked with cynicism if their hard-earned results suddenly roll out of another laboratory.

There is no need, however, to debate the merits of secrecy when a discovery is so all-embracing that it sews up a field. Then you have nothing to lose by immediately shouting it to the world. The double helix that Francis Crick and I came up with in the spring of 1953 was such a case. So naturally we broadcast it as fast as possible, knowing that if we would wait, someone else would inevitably think out the right answer and we would have to share credit. Desires to ward off competition, however, may also tempt you to make claims beyond your data, with the thought that others will give up the chase and for a few months let you do science at a more civilized pace. Assertions are often made informally that so-and-so has been tightly established, say for example the purification of a key enzyme then at the center of biochemistry. Receipt of such news might induce you to give up a similar project, since it would be criminal to keep a younger colleague on a project that already has been accomplished. Only months later do you realize that the key experiments had never been done correctly and that the problem is still for picking. So a discerning eye is often needed to know which bits of unpublished information should be taken seriously.

A clear head is also needed when you hit upon a crucial technique or observation that immediately will make possible scores of other important experiments. If you already possess a large lab with many competent students, then you can have the joy of following up your discovery to its natural conclusion. But if you are unknown and with little resources at your disposal, you may be certain that if you talk openly other labs will rush in and you may quickly be frozen out of a field that owes its existence to your ingenuity. The only sure way to prevent such a disaster is to let no one outside your lab onto your secret until you have exploited all its obvious consequences. Then you can have the pleasure of a public announcement, uncomplicated by qualms that you are creating unwanted competition. Of course, it may be very tricky not to let your exuberance betray that you have struck oil, and rumors may fly that you have a big secret. Some people may be upset, but usually only because they figure you must have let your close friends in on your act and they feel discriminated against. Once, however, you reveal what is up, the beauty of your results will soon make most everyone forget your long period of silence.

In contrast, your reputation is likely to suffer if you make a stink about unwanted competition. Your original generosity in letting everyone onto your big secret early could easily be forgotten if you want to eat your cake and then have it. There is no unwritten law among scientists that given problems belong to specific individuals, and agreements among peers on how to divide a tasty pie usually lead to further anguish when it becomes apparent that one side has got the better bargain. To be sure, special pleading may keep some compassionate rival from taking up the scent for a few months, but if what you have announced is at all revolutionary, you must expect the deluge, if not from one side, most certainly from the other. So once your news is out, the only sane course to follow is to work like the devil, hoping that your incipient competition may take several months to get into the act.

Aside from deferences owing to close friendships, which can extend over national boundaries, the only rule that most people follow is the avoidance of problems that someone in their own environment has already latched onto. Only saint-like minds can watch someone in the next lab race them for an experimental result, and not get violently upset. Even when minor scientific points are involved, someone's temper will give way, and if friendship ever existed, it will soon turn to antipathy, if not hatred. While competition hundreds of miles away need not upset your day-by-day existence, when it is next door it becomes a canker the brain cannot ignore. This hard fact of scientific life means you may be in for real torture if someone in your own university, if not department, makes an unexpected observation that you would like to exploit. You know that if you were at another university, you would have no hesitation in jumping in on the fun, but if you must coexist within the same faculty, you should tolerantly grin and hope that after the first flush of excitement passes, there may be an oblique way to get into the action.

To be sure, I know of people who have not followed that advice, usually on the false assumption that they have been begotten with talents not given to those who started the ball rolling. Such bully-like behavior however is difficult to pull off, and even when the most insensitive of egotists move in, the storms they ignite are usually more than they have bargained for. If not stopped, they can often paralyze an institution, and those at the top have no recourse but to tell the offending characters to back down or get out. If cooperation instead of competition were to emerge, everyone would breathe relief, but if someone wants to do his own thing, he should have his way. Of course, there are many cases when a lone wolf would have been very wise to get more talent on his side and suffered badly for trying experiments over his head. Much depends upon how help is offered, as no one ever feels happy that his own small operation has become part of a large corporation over which he has no effective control. Owning shares in a conglomerate may make sense if you sell out at the right time, but all too often they are not worth the paper they have been printed upon.

Always a key question is the number of the people who optimally should be in a field at a given time. All of us would be very bored if we did not have the results of others to look forward to, even within our own highly specialized area. I know of no one who can generate anywhere near enough unexpected new facts to provide internal contentment. New issues of our scientific journals are all too often needed to knock us out of the lethargic grip of our own thoughts. And though we may occasionally bitch that there are too many scientific meetings, they are generally indispensable for providing advance clues as to what will come next. Your pulse, of course, begins to skip when you see a title that might imply a solution to years of your research life, and what a relief it is to generally discover that someone else also does not know what is up.

So sanity generally demands some undefinable balance between your productivity and that of the outside. As long as you can continue to turn out good ideas, you can take joy in the success of others and actually promote their careers as essential to your own. But if you are going through lean years and see your research support dwindling, the cheerful arrogance that characterizes tough science on the move may look increasingly indistinguishable from the ethic of the business bastards who worry about human frailty only as it affects profits. The love affair with the scientific life that so dominates our lives as young adults is thus not easy to maintain, and, like a real marriage, will most certainly collapse if you expect too much of your love object. Unfortunately, the treatment of the scientific life fed to us as children bears little relation to reality, and the glowing portraits of our selfless characters that dominate virtually all popular books about science read as if cribbed from the lives of the unmartyred saints. So the fact that scientists are no better or worse than others of our backgrounds often is not learned in time to prevent much unnecessary disillusionment.

Care especially must be taken not to credit scientists whose research bears directly on human problems as less liable to human temptation than those who fix on areas lacking immediate relevance to mankind. The press releases that tell us of vast collaborative efforts to understand horrific diseases or to bring forth necessary new forms of energy all too easily can be read to mean that we are witnessing the births of new forms of human cooperation that will light the way for future decades. My experience with such programs, however, is that this is all hogwash. Such programs often get going not because they make scientific sense but because the public is crying for relief and politicians do not like to admit they are impotent to help their constituents. So, independent of whether we have a fighting chance, we can all too easily charge ahead, say against the boll weevil, and in the long term only make some cotton congressman happy, by providing soft jobs for his less accomplished relatives.

As a result, scientists who, for better or worse, get involved in such applied projects, usually have poorer prospects than those whose career choices were essentially intellectual. They know that unless they rapidly move toward their goals, they will never make it big, much less keep their often very well funded jobs. For, usually, they are backed by "soft" monies that suddenly might disappear, and if progress is not repeatedly proclaimed, the ax will fall to allow some new pressure group to have its chance. So the atmosphere aboard these massive national efforts can be most tempestuous, with conflicting approaches subject to great skepticism if not scorn by their respective opponents, and endless jockeying for next year's appropriations being a major aspect of daily life. Such programs thus become naturals for scientific messiahs, for with charisma at the top, the getting of next year's budget becomes a much more certain affair.

Now of course there are many problems where science may have a real chance to help mankind. Here you may ask whether a cause can ever be good enough to generate small worlds of selfless scientists where the common good is the overwhelming consideration and thoughts about priorities receive little attention. In fact, don't we already have such an example in the recent "Conquest of Cancer" legislation? It not only provides to cancer research resources unimaginable several years ago, but also tries to specify how dispassionate advisers can oversee its swift running. I'm afraid, however, here again we have an unrealistic pipe dream and that in so far as we are dealing with the process of scientific discovery, the sociology of cancer research will not show any striking differences from other branches of science. Even if we disregard the tons of lousy cancer research still being done and concentrate on the first-rate variety, which for the first time is beginning to roll out in large masses, we still must deal with the usual complicated mix of cooperation and competition.

Again, there does not exist any practical way to insure that two talented people never want to solve the same problem. At a given moment, certain areas will look more glamorous and about to crack, and I see no practical way to push first-rate minds into less attractive approaches. Moreover, some duplication of effort may be the only way to reach a goal within a certain time span. No two minds ever take exactly the same path, and placing all your marbles on one person's intuition never makes sense if you have the money and want to move quickly.

Though the National Cancer Institute has many large-scale projects set up specifically to bring people together so that they can pool their data, not all labs want to operate that way. I even know of first-rate labs which occasionally operate on the "need to know" procedure favored by government security agents anxious to keep key information from getting to inappropriate offices both within and without our national government. Scientists may only be let in on what their colleagues in the next room are doing when such facts become necessary to let them proceed onto the next crucial experiment. Such behavior, of course, is at times a consequence of fears that someone will use your preliminary results to beat you to the punch. But it may also aim to prevent the spread of hot facts that cannot later be substantiated. No one may take you seriously if one day you appear to say one thing, but the next week blurt out the opposite. Sticklers for not making mistakes often compulsively resist letting out facts which cannot be backed up with a written manuscript. Irrespective of its origin, however, secretive science has the great limitation that you often do not try out your ideas on your friends to see if they can spot a flaw that you may have missed. So, unless you are very good, mistakes can go on longer than necessary and the quality of your science goes down.

The question thus becomes posed whether obsessive secrecy is delaying the day when we can at last understand the molecular basis of cancer. If so, the public has every right to feel betrayed with its so far unwavering support for cancer research degenerating into the cynical attitude now directed toward almost anyone who makes a business of working to better the human condition. My guess, however, is that, at worst, these manifestations of the secretive urge will have only a marginal effect on the final result. On the one hand, secrecy obviously harms when it keeps people banging away on unproductive leads for longer than necessary. On the other hand, perhaps an equal number of scientists would lose their effectiveness if they did not have the peace of mind produced by limited secrecy. Unless you have unharried time to reach conclusions by yourself, unhampered by the thought that the next minute will bring a telephone call from someone doing the same thing, you can never acquire the self-confidence which comes from the realization that you can think through difficult problems or have the persistence to stick with an experiment until it is solved. Some forms of solitude are necessary for self-identification, and we would seriously misjudge the human temperament to suggest that we always work best in the company of others. Of course, in many instances teams of two or three work well together, but they're often most effective if they collect different skills and members are not interchangeable.

A natural limitation thus exists on the numbers of people who can effectively cooperate with each other; once groups get bigger than a certain size, they tend to fragment. Two is the ideal number for most projects, since you always have a partner to keep you going when your morale temporarily lapses, while there is no opportunity to be the odd man out when your talents are no longer needed. Very difficult projects nonetheless may only be attackable if many more hands are available, but even so, large groups will stay together only as long as each member feels unique and not used by the others. Within the same walls, it never makes sense to have too many people with the same general objectives. As long as the expenses incurred in the duplication of facilities do not overwhelm us, it is instead better to set up a large number of different research groups working in different spots. This choice, of course, leads to the specter of duplication, but the advantages of feeling that you are master of your own fate will more than compensate for the tensions generated by the resulting competition.

Frequently, of course, you find that you can not only learn from, but also like your competitor, and if you meet often enough at meetings and through visits to each other's labs, the close friendships which develop will lead, if not to jointly conceived experiments, most certainly to agreements to publish your data simultaneously. The fact that one of you may have reached a conclusion several weeks earlier matters much less than the fact that both of you had the sense to work out ways that would eventually culminate in the desired answer. Unfortunately, such gentlemanly agreements often pan out better when the objective is minor and not likely to attract many of your peers' attention. When you suspect that the correct answer might spectacularly alter current thought, the temptation is bound to arise to question the desirability of sharing rewards with someone with seemingly less credentials, particularly if he has a long history of jumping in on hot problems that only the hard work of others has made capable of quick solution. So, while many highly compassionate individuals exist and never behave in a dog-eat-dog fashion, the exact behavior we can expect from most colleagues is likely to depend upon the particular situation and, more than often, passions will temporarily flare up and curse words be the temper of the moment.

So I see no way that the rat-race aspects of much of high-powered science can vanish in the foreseeable future. To think otherwise would be to go back to the utopian ideals of the commune, a concept almost as old as modern man and whose periodic revival leads always to the rediscovery that selfishness often wins over altruism. But this realization does not mean we should be tempted to give up the scientific life. Initially, we chose it because the unknown in science was irresistible, and, more than likely, in other worlds we would be not only out of place but very bored. Moreover, other realistic career choices would not allow us to escape the competitive pressures of the modern world. No matter where you land, you have to do something well, and that will never be easy. Each of us must thus somehow put our scientific dreams in line with the balance sheet of our limitations and capabilities. For if we do, we may perpetuate the excitement of our youths and still be able to anticipate each new day as a unique opportunity to follow our heart's desire.

CREATIVITY AND
COLLABORATION IN
TELEVISION PROGRAMS

HUW WHELDON *A creative force behind the coming of age of the documentary in British television, Welsh-born Huw Wheldon is now Managing Director of Television for the British Broadcasting Corporation.*

Wheldon joined the BBC in 1952, making his mark early as producer of award-winning "PORTRAITS OF POWER," "ORSON WELLES SKETCHBOOK," and "EPIC BATTLES." Wheldon went on to become first producer of "MONITOR," the BBC's popular bi-weekly program on the arts. Maintaining his ties with "MONITOR," which he frequently narrated, Wheldon rose to be Assistant Head of Talks in 1962, Head of Documentary Programmes in 1963, and shortly thereafter in charge of Documentary and Music Programmes. Wheldon became Controller of Television Programmes in 1965, a position he held until 1969, when appointed to his present post.

IF THERE IS a sentence I dislike more than "Television is, of course, an art form," it is "Television is, of course, not an art form." All in all, I feel I have to move with some circumspection when it comes to words like "creativity." I flinch from them. Equally, I am suspicious of the word "collaborator." I tend to pronounce it "collaborateur." Hobnobbing with the enemy. At the same time, apart from being keenly conscious of the honor done me and done the BBC by this invitation, I readily acknowledge that the theme is important to the contemporary world and to television in it.

I propose to limit myself expressly to various practices and methods which have been evolved over many years in the BBC and which seem relevant to this theme; and in doing this, I must emphasize that I am not suggesting the applicability of such practices elsewhere, least of all in this country. The BBC is a highly idiosyncratic organization, very sophisticated and as British as Hollywood is, or at any rate was, American. You do things your way. We do things in ours.

Let us start then by a brief consideration of a few BBC Television programs. I have tried to choose programs which can straightforwardly be described as creations and to eschew programs which can equally straightforwardly be recognized as potboilers. *Civilisation* was a series of thirteen fifty-minute programs which we made on locations across the world. The series was originally started by a group of BBC program executives, of whom I was one. What we wanted was work with and from Kenneth Clark. We did not know how many programs we might get made, nor on what theme, nor what form they might take. We were trying to establish a second BBC television network at the time. We were making *The Forsyte Saga* to go out on the new network; we had made some hair-raising comedies, including a series with Dudley Moore and Peter Cook; and we wanted to make the best of our opportunity to break new ground in the documentary field. So we went for Clark.

As we discussed possibilities, we began to feel that the project as a whole should be at least the equivalent of an important book on a significant subject by an author who cut ice. Ideally, it should be describable as "an original work." I was, myself, clear that Clark must be an organic part of the production team from beginning to end, not a visiting fireman putting a gloss here or making a flourish there. Least of all did we want him to appear in the programs as a celebrity. Programs, like other works, are made in the making; and Clark had to be part of the making. He agreed to all this, and we finally decided on the series you know about. Budgets, lengths of programs, and timetables were set up, and staffs and facilities were assigned. The original group of executives dropped out other than to be there if required. From the beginning, with much profitable disagreement, it took about a year to arrange and over another two years to make. Clark was at one with the production team throughout, roaming the world with the rest of them, making the series.

The creative impulse in the matter of theme came from Lord Clark. His mark was upon it. The series was about Western civilization, but by the selections he made, the tone he adopted, the emphases and ironies of both word and image, there was a valid sense in which the work as a whole constituted a personal essay. It emerged out of a particular intelligence and was inseparable from a particular sensibility. *Civilisation* was about civilization; and it was also about Kenneth Clark.

The man who made the programs was Michael Gill, one of my colleagues, a member of the BBC staff; in our usage, the producer. No one who knows his work will deny that his stamp was upon it, too. Those who saw the Alistair Cooke *America* programs, which Gill produced later, will recognize that both series had something in common, an idiosyncrasy of style, again coming out of a particular intelligence and inseparable from a particular sensibility. What they had in common was Michael Gill.

The extent to which Clark drew upon Gill, and Gill drew upon Clark, is not open to measurement or even to speculation. They had separate responsibilities, but they worked together. What is clear is that, between them, they provided the fundamental creative impulse upon which the series was built. Kenneth Clark and Michael Gill constituted a successful collaboration; but as with other great creative partnerships, they also depended upon an assembly of many separate skills to bring the whole thing to fruition. They could not possibly have managed without the help of a small army of technicians and accountants and electronic engineers and cameramen; and in this sense, television, like all the performing arts, is necessarily deeply collaborative. Many skills are called on, and it is very much part of the creative capacity of the producer to be able to guide, teach, lead, and inspire all concerned to play their parts to the full so that what he has launched can reach its true destination. But the primary impulse is what here concerns us. That central thrust does not necessarily, of course, involve a partnership.

Till Death Us Do Part, to take a comedy series, is also, I think, genuinely a creation, and depends very much, as it happens, on the work of one man: its author, Johnny Speight. CBS's *All in the Family* is based on this program. Where *Civilisation* was long, speculative, and factual, *Till Death* is short, explosive, and funny. Both of them, however, create an original world. Speight has been brilliantly served by his cast and producer; but it is his mark—the author's—that is on the program.

It is also true that sometimes the main creator is what we call the producer. *The Search for the Nile* (I choose it because it was also widely seen in this country) was a project pressed upon us by Christopher Ralling, who, as a documentary producer, has been on the BBC staff for many years and who, for almost as long, has wanted to do this series. In the event, program heads recognized, if not his idea, then Ralling's right as a proven maker of good documentary programs, to be supported, and the series went ahead.

I make so bold as to remind you that there is no such thing, of course, as "a good idea for a program," any more than there is a good idea for a poem or a novel or a play. There are only poets and novelists and playwrights. In this particular event, Ralling researched it, wrote it, directed some of the episodes, and, as its producer, was in charge of all of it. It was his stamp that was upon it.

On the whole, however, and in my experience, the main creative impulse in most television programs comes from a combination of two or three people. Occasionally, there may be one or two more added. It is certainly true of the plays we produce. Both author and producer are always involved, whether we are talking about plays by Ibsen or Chekhov or about plays by J. B. Priestley or David Mercer. It goes without saying that Christopher Morahan and

Alan Cooke contributed more than just organization and camera direction to Pinter's *A Night Out* and John Osborne's *Luther,* as did Basil Coleman to *The Tempest* and Michael Elliot to *She Stoops to Conquer.*

In some ways, I would prefer it if the individual act of creative leadership came out less strongly. There is something agreeable as well as fashionable about the idea of a group of people out of which creation emerges, each member equally participating, none of them in the limelight, but all acting modestly and creatively together. That is attractive. I have no doubt, it occasionally takes place. We have ourselves mounted plays which came out of the deliberate and conjoint effort of actors and production people and the surrounding community, and in which the presence of the writer was more editorial than epistolary. We know that the translation of the King James Bible really did emerge out of the work of a marvelous committee. It seems more than possible that some of the temples and cathedrals and cloud-capped towers of the world really did soar into the sky through the conjoint collaborative work of dozens or scores of artists and craftsmen bound together by a great purpose and a common culture. On the whole, however, I beg leave to infer the presence of a masterbuilder somewhere in the neighborhood of most edifices which glorify the world, be they of stone or of words. I also beg leave to infer a general absence of committees or panels in the creation of any singular vision.

Among the short and merry programs which seem to me to be genuine creations is one we make in the BBC called *Steptoe and Son.* Like *Till Death Us Do Part,* it is wildly popular in the United Kingdom and is seen by over a third of the entire population whenever we put it out. *Steptoe,* again like *Till Death,* is one of those creations which draws its power as a work primarily from its author. As with most of our comedy shows, it is again brilliantly served by its cast and its producers; but the stamp that is on it seems to be pre-eminently the author's. In the case of *Steptoe and Son*, the author is two people, Ray Galton and Alan Simpson.

On the surface of things, *Sanford and Son* is the American version of *Steptoe and Son*, just as *All in the Family* is the American version of *Till Death Us Do Part*. I say "on the surface" because, in fact, the two groups of programs serve different purposes and are based on different and separate traditions. *Sanford and Son* and *All in the Family* are seen every week, month in month out. By today, the episodes in each series already transmitted in the United States must run into three figures. *Steptoe and Son*, on the other hand, has been running irregularly in the United Kingdom over a period of twelve years, and, during the whole of that time, Galton and Simpson, working on commission from the BBC, have written only fifty episodes all together, something less than five episodes a year. Johnny Speight, equally, has written thirty-nine episodes of *Till Death Us Do Part* in eight years. I could have wished, with both series, as with others, that there had been many more. One a week would have been ideal. But four or five short pieces was what the authors concerned felt they could write on a year's average and still keep the works fresh and truthful. That is what they had it in them to write, and we have had to be satisfied with that.

The great creations of English literature have, I suppose, always been too few for their fans. I could have done with many more appearances on the part of Mr. Micawber or the Red Queen. Even when they did appear again, quite often, as in the case of Evelyn Waugh's Basil Seal or Wodehouse's Jeeves, they were still too infrequent for me. Conan Doyle could have written Sherlock Holmes stories forever, so far as his readers were concerned. He finally wrote sixty in all, including four novels. That was what he had it in him to do.

Steptoe and son have so far appeared in nearly as many situations as did the great Sherlock Holmes. Nor is it inappropriate to mention Baker Street and Latimer Road, West London, the home of the Steptoes, in the same breath. Situation comedy in the United Kingdom owes much to actors and is deeply lodged in the popular literary traditions of England.

Why it is that British actors are so good is a mystery. In particular, why it is that English actors are so convincing is baffling. I speak as a Welshman, and can only suppose that some national and psychic compensation is at work making up for stiff upper lips elsewhere. If you want to cast a play in English, my advice is as follows: dismiss all actors and actresses called MacDonald or Stevenson; reject everybody called O'Brien or Murphy; or Kreisky or Pollecoff; or Jones or Williams; and simply take on people who carry those repellent Anglo-Saxon surnames which sound like fresh-water fish: people called Tench and Carp and Chubb and Pike and Dench and Bream and Bleak. Plays with John Tench and Dorothy Bleak in them will be splendidly acted. Whatever the reason, English actors are, in fact, brilliant; and, as the saying goes, they grow on trees.

Situation comedies in the United Kingdom, then, are very much written by writers and acted by actors. Sometimes these comedies are good, sometimes indifferent, but for better or worse, the literary and dramatic tradition of England is what they belong to. It is one of the great traditions of the Western world. For that reason, we seldom keep, or indeed can keep, a situation comedy show going for more than a few episodes at a time, for precisely the same reason as publishers can rarely bring out more than one novel or new collection of short stories every year or two from the same writer, no matter how fruitful he is. It is a question of what is humanly possible. It is infuriating, because our demands know no bounds; but there it is. If you want things to be written by writers, you have to give them time and accept limits.

Situation comedy in America has a different tradition. It was here, after all, that it was invented, in the early days of radio. A comedian in the old vaudeville-circuit days could live out a lifetime on the equivalent of three or four hours of material. Broadcasting put an end to that. Once used, the material was finished. So the idea of setting up domestic or social situations to serve as vehicles for comedians was invented. The element of creativity in such shows in the United States seems to me to lie more in the performance and in some degree in the production than in the writing, and bearing in mind its comedian tradition, this is not surprising. Of all the many programs of all sorts we have bought from the United States over the years and shown on BBC Television, the ones I personally miss most are the Phil Silvers *Sergeant Bilko* show and the original Dick Van Dyke show. They seemed to me to be genuine creations, although they were turned out week after week as if on a conveyor belt. Nevertheless, they looked handmade and not merely manufactured.

I am not claiming that either of them was a great cry of the human spirit, but each was a real piece of work and not a potboiler. *Sanford and Son*, like *All in the Family*, seen week in week out with never a pause, you may think take their place in the same league and in this same tradition of production and performance. Be that as it may, the names of their authors in the case of any of these programs will not spring to mind. They are not, strictly speaking, the true equivalents of *Steptoe and Son* or *Dad's Army* or *Till*

Death Us Do Part or *Not in Front of the Children,* which belong to a different set of traditions.

The names of their authors (unless he was Orson Welles) do not spring to mind when one considers very good American films, either. Their thrust and power also seem to reside in performance and production or direction, rather than in authorship. And if this is so, how to define that particular tradition of creativity and having defined it, how to evaluate it and weigh it, and then how to establish and develop it in other forms of communication including television is clearly for others and not for me to think about.

So far as the BBC is concerned, there is no doubt that the creativity of its programs lies firmly in the literary and dramatic traditions of the country. We feel that, like the theater at large, we should be wanting if we did not ceaselessly recreate the classics: Shakespeare, Sheridan, Shaw, and so on. We also feel that we would be wanting if we failed to commission plays for television from the contemporary playwrights of the kingdom; and we do so throughout the year. It is equally significant that whether we move away from the central concept of the play to situation comedies on the one hand or to documentaries and features on the other, it is, more often than not, to individual producers working with individual writers and to individual writers working with individual producers that we normally turn. The degree to which we can attract first-rate minds into television programs as producers or as writers is obviously crucial. What is no less crucial is the degree to which this turns on independence.

Before I turn to the subject of independence and creativity, let me touch upon programs that are not "creations" in the sense I have been using the word creativity.

A quiz program, for example, or the coverage of baseball

or football certainly calls for skill but, simple or complex, does not call for the degree or kind of inwardness which characterizes an original work. A whole range of important programs which can accurately be called journalism are not, strictly speaking, creative in the sense in which I have been using the word. At the same time, all programs can and should be well arranged. Ordinary and even simple programs provide most, if not all of us, with true, cheerful, and sometimes intense pleasure. If they are made badly, they will only succeed in degrading what they are meant to serve, including the audience.

The quality many such programs turn on is not so much creativity as judgment. They need flair and skill, but, above all, they need judgment. I believe that the various judgments which have to be made in connection with such programs are, as it happens, much helped in the BBC by having to be formed within what is an intensely critical, creative, and collaborative community.

I ought to interject here that BBC Television, and in this respect it is crucially different from the great American networks, itself makes over 85 per cent of all the programs it transmits, not counting movies. We run two full national networks reaching out to every part of the country, and, broadly speaking, if there is fiction on one there will be fact on the other; if a documentary or sport on one, a movie or play on the other. We ourselves make, apart from what we buy, over ten thousand hours of programs a year.

BBC Television is largely concentrated in Shepherds Bush, London, and there seven thousand of us cluster around the studios and sets and cutting rooms and stages. The place seethes with producers and designers, librarians and cameramen, planners and carpenters, film editors and papier-mâché workers. We are joined daily by further thousands who are not on the BBC staff as such, but who are with us temporarily for given programs: actors, orchestral players, comedians, dancers, pop groups, choreographers, scholars, writers, politicians. What we make up between us is willy-nilly a creative and collaborative community.

To return to my theme of creativity and judgment: if judgments about programs that have to be made every day inside that community are helped by the prevailing climate of creativity, still more, I think, is creativity itself helped by a prevailing climate in which true judgments are expected to be made openly and frequently.

The most important standard of judgment and the one subsuming all others is truth; that is, the distinction between what is true and what is false. For the past fifty years at the BBC, the one condition necessary for truth to be told has been daily, sedulously fought for, defended, and nurtured. That necessary condition is the independence of the BBC of all other interests, be they governmental, commercial, political, educational, or belonging to any other dominion or principality. One of the characteristics of the BBC as a creative community is its awareness of this daily fight. Independence must be defended if the truth is to be told. I will return to this point later.

I hesitate to use the word "creative" in connection with news, because I would not give the smallest hostage to those who would bring views or opinions into news bulletins. BBC news has always had great authority. People on the whole believe that if a news item is "on the BBC" it must be true. The authority of the BBC in the United Kingdom is greater than that of any other medium of communication. It is a magnificent reputation to have, and it is one which we enjoy in some degree abroad as well as at home. It is a difficult one to hold on to in an increasingly divided and dividing world. What it is based on is a simple and laborious preoccupation with the accuracy of fact as observed, reported upon, or recorded. Accuracy rests on habits of check and countercheck; on independent judgment, independent observation, and on absolutely scrupulous shooting and writing.

It is an interesting technical fact that in the BBC we separate news reporting very strictly from public-affairs programs. No news reader makes a comment or takes a view on any item of news, ever. We are careful to the point of pedantry. The staffs, cameramen, editors, assistant editors, everybody connected with news, do not, generally speaking, work on other programs.

The truth of a news bulletin is the degree to which it accurately and independently describes an incident which has taken place. The truth of a play is not so different as might at first appear; the truth of a play or a creative work is the degree to which a world conceived inwardly has been accurately and independently embodied forth; the degree to which it is not meretricious, the degree to which it hangs together as a single object, cutting no corners, cheating no one, including the author. What inwardness is to plays, and accuracy is to news bulletins, disinterestedness is to public-affairs programs and to certain forms of documentary. Pretentiousness, in my own experience, or merely fashionable attitudes are what serve most frequently to diminish plays, just as a note of self-righteousness and facile assumption is what undoes public-affairs programs on the occasions when they are less good than they should be. In my view, there are more real potboilers and cardboard cutouts among so-called "serious" programs than there are among so-called "ordinary" programs. In any event, programs which emerge from inwardness, and programs which are based upon the judgments forced upon us day after day, and programs, finally, which need to be disinterested, that is, taking in all factors and being above the battle, will equally collapse unless they are founded in the truth.

Let me turn, for a moment, more expressly to collaboration. I will not here dwell on the collaboration that must exist between, for example, a director and a writer or between a choreographer and a designer or between an editor and a producer and a writer or between a director and his cast. I have had direct experience of most of these associations, but they are neither new nor unfamiliar. They did not come into the world with television, and they have been explored over and over again by people as well or better qualified than me to explore those areas of flashpoint and dynamism. They are not new, and existed long before broadcasting. What is new is television itself and the degree to which it is asked or required to collaborate with other forces in society. Television is technically so laborious and so expensive that it has had to turn, in various parts of the world, either to government or to commerce for the enormous investments it requires in order to operate at all; and governments and commerce have accordingly called the tune. In some respects, the pressures have been toward collaboration and away from creativity.

I will not waste time on the dangers of collaboration with government. They are sufficiently obvious; what comes out of government broadcasting services is at worst propaganda and manipulation and, at best, official communications, official art, and officialese generally. All of them are spiritually deadening. People in America always think that the BBC is run by the government. It is not so. We believe we have evolved a system of broadcasting that has kept the government's fingers out of programs and programs out of the clutches of the British equivalent of Madison Avenue. We may be deluding ourselves. I think not. The arrangements are sophisticated and fit nicely within the quite surprising complexities of our constitutional system. Neither chairman nor board member, neither senior nor junior member of the executive or production staff has ever changed with a change of government. No BBC program has ever been taken off or put on either directly or indirectly by government agency. No BBC program has ever been sponsored by anybody. We get the money we spend, the equivalent of something over three hundred million dollars a year, from the households that have sets on which they can watch our programs. There are no debentures or loans or shares or anything else. We finance our entire activity, including all our capital expenditure, out of income. It is a kind of subscription television based upon a single annual payment. The Post Office collects the money for us, and we pay them twenty-five million dollars a year for doing so. We could collect the three hundred million ourselves if we wanted to, but we calculate that it would cost us even more.

Far from being subsidized by the government (or by anyone else), we have to pay taxes where they apply, like any other public corporation. We used to be required to pay income tax, but we took the government to the High Court on that, and beat them. Even so, in the postwar period we have paid over to the government by way of tax not less than one hundred and twenty million pounds; and that is a lot of dollars. So much for the government subsidies!

Independence from government is crucial, if creativity is to exist at all. But commercial pressures work in more mysterious ways their wonders to perform. Gone or going, of course, in most parts of the world, are the days when television programs were simply extended advertisements, creatures of the sponsor. We have in our country, and entirely separate from the BBC, a form of commercial television in which the advertisements are completely distinct from the programs, and sponsoring as such is actually illegal. You can buy times, at different prices, for different parts of the day, for your commercials, and that is that. By commercial pressure I am referring to something altogether more insidious and to something which bears down upon the BBC as it does upon other broadcasting organizations, commercial or not. I mean commercial habits and the commercial ethos. Let me touch upon an aspect of "productivity," itself a word imported into television from the world of industry; and upon the concept of the package, equally something imported into the world of television from the world of commerce. No genuine creation has ever been conceived as a "package." I mentioned Sherlock Holmes. You will remember that Raymond Chandler wrote, I think, only seven novels altogether. He must have exasperated his publishers. They must have wanted more. On the other hand, they settled for seven because that was all there was.

Television companies have done the same. They have had to settle for what was available. I cannot help speculating, however, that, had Raymond Chandler lived a little later, the television world might have been more greedy. The equivalent of seven novels would not have been enough. I can very easily imagine television executives here or in any part of the world setting up a Philip Marlowe format onto which writers could be assigned and out of which they could confidently expect not a handful, but hundreds of episodes running over years. Nor would that be thought to be unusual. And yet it is to exchange the true creativity of Raymond Chandler stories for the altogether different quality of a Raymond Chandler-type series. What is lost is precisely the true vision of that small world he made, which belongs to Chandler himself. In an exact sense, it is to offer an ersatz product. Form has become format, and content has yielded to calculations about contentment.

The practice of enlisting a team of writers to work together to write gags for a comedian is one thing; to assign a group of writers to work together on a narrative script is another. Properly speaking, this is copy writing. It is the way advertisements are written. You get exact, sophisticated, and brilliant work which makes an immediate impact for a precise purpose. What you do not get and what such a procedure cannot give you is the inwardness of an individual creation. Copy writing is both efficient and reputable in its place, but it is not literature and it is not art. It is tempting, however, to television organizations, and it tempts in the name of productivity.

The stakes, in the context of the British experience at all events, seem to be very high. We look back on the nineteenth century, see its achievements, and regret in many ways the passing of some of its pieties. We also cannot avoid noticing its slums. They constitute an unquestioned acceptance of the proposition that there had to be whole streets and neighborhoods made up expressly of second-rate dwellings. A whole nineteenth-century culture decided that in one of its aspects it was desirable and sensible or anyway unavoidable to commit itself to mediocrity. It is mediocrity, the cracked and dirty pavements of George Orwell's terrible city, that is the great enemy of good television; and potboilers and packages and programs calculated rather than made, turned out rather than created (with whatever measure of success or failure) are mediocre by definition. They are slums of the spirit and slums of the mind.

For all that, with many a rationalization, many a resounding and false justification, we sometimes fall into that trap. We swap created quality for uncreated quantity.

When that happens, we are settling for cardboard as against art; and in that particular equation you have to plump even for third-rate art as against first-rate cardboard. It is all too easy, in the name of being "businesslike" and effective, to get trapped into the world of the slum.

One of the more promising collaborative developments of the past few years goes under the umbrella title of co-production. Co-production can mean at least two sharply different methods of collaboration. There is, in the first place, the kind of arrangement whereby two or more organizations, generally broadcasting companies, combine on a single project, each calling upon the other or others to provide different and separate services or production skills. Thus, the BBC worked closely with ABC of Australia in broadcasting Prokofiev's opera *War and Peace* when the Sydney Opera House was opened. We provided the production skills, and ABC provided the services and facilities to the requirements of our producer. Equally, we worked closely with CBC of Canada on *The Wit and World of George Bernard Shaw*. In this case, they provided the producer and we provided the facilities. Each production was firmly in the creative hands of an individual producer from one or other of the organizations committed to the programs.

The other, quite different form of co-production, and this applies, as it happens, to most of the programs I have mentioned earlier, is where one organization invests money in a program or series of programs wholly made by another. Thus, in the case of Alistair Cooke's *America* series, for example; or the Jacob Bronowski series *The Ascent of Man*; or *War and Peace*; or its successor, the Trollope novels, *The Pallisers*; or *Search for the Nile*; money was invested in each case by Time/Life, who are indeed our major partners in these matters in the United States. From our point of view, it has meant getting more money to spend on already expensive programs, and has therefore been much to our advantage. From the point of view of Time/Life, it has meant being associated with programs which have, I hope, contributed to television in America, and which, I also hope, have brought them prestige and honor as well as satisfactory returns on investments.

We have entered into various relations of this kind in different parts of the world. They vary, of course, in detail as between one and the other. I must make plain, however, a set of presuppositions which are general to each and every co-production everywhere. We will not accept any strings where production is concerned. We are adamant that final editorial control is ours, absolutely. To make sure, in this naughty world, we write this into the contract; and by the same token, we insist that the major financial investment comes from ourselves, always. He who pays the piper calls the tune. All in all, we are nothing if not stiff-necked. We expect other broadcasting organizations to behave in the same way. I am concerned here with a general principle, not with a BBC convention.

The general principle is that difficulties arise if there is the opportunity to interfere with the central creative force. An organization which has invested money or facilities in a production could be forgiven for wanting to be in a position to influence the story or the style, to make helpful suggestions about casting or about locations. And yet, if my central propositions and my general theme are a true bill, if there is one thing that has to be placed, for better or for worse, in the hands of the individual program maker, it is precisely the cluster of decisions that is summed up in the words "story," "casting," "location," and "style." Collaboration has its possibilities and they are real. It also has its dangers.

In a word, beware of agencies or agents that want to help you make television programs. They have their own purposes and they will not always be consonant with yours. What we need are dynamisms or organizations that are rich enough and strong enough and civilized enough and independent enough to see all productions, including co-productions, as turning on the creativity of individually responsible program makers. We have, in fact, after a stuttering start, suffered no serious difficulties in the matter of co-production. Time/Life in particular, as you would expect of an organization long accustomed to writers and writing, have been brilliantly rewarding partners. On the other hand, we have been, as you would expect, very careful. We are able to be so, because we have inherited altogether singular privileges at the BBC. The charter and license foundation is in fact an extraordinary constitutional invention. It imposes upon us the duty to take programs seriously, and purely for their own sakes. We are encouraged and empowered to make the best programs we possibly can. There is nothing, neither government nor commerce nor any other force, to deflect us from that aim or to press us toward compromise. If we do badly, and we do so all too frequently, it is firstly because any organization that is seriously committed to genuine creativity must have the nerve to accept a failure rate; and secondly because of our own foolishness and frailty, our own failures in skill, or care or thought. We are encouraged by the very system to make plays and operas and original works. At the same time, we cannot hand tablets down from Mount Sinai. The same charter and license foundation insists upon a constant reference to the audience, because it is they who pay. In a word, we are protected from inhabiting ivory towers or slums. If we settle for the pretentious or the false on the one hand, or for the potboiler or the "product" on the other, it is because we fail to use our own foundation to proper advantage. Indeed, in so doing we only arm our enemies and put at risk an organization which, whatever else it has collectively learned over the years, has learned that creativity is truly individual and must be protected, and for that, independence is essential.

In fact, the whole system I have here touched upon depends upon the collaboration of all creative people concerned in this daily enterprise of fighting to keep the BBC's independence. The success of the enterprise depends on the members of that community understanding the relationship between their own freedom and the BBC's independence. They are free to do what is true for them and what can be defended, in the common language, with true conscience. This, I think, is the chief collaborative characteristic of the BBC as a creative community. It becomes even more essential to create or to protect organizations or principles of action which can cherish individual creation as we move into an increasingly technical world. Thinkers not given to hollow phrases and as far apart as Walter Benjamin and F. R. Leavis have noted that as information and communication have made their way, so creativity and imagination have fallen back. I have myself over the years argued that the story as such is the most potent discovery, of a way by which we can examine ourselves and examine the world around us, yet made by men. It is overwhelmingly through stories that we learn what we are like, and it is through stories that we learn what the world is like to live in. It is no accident that religions teach by parables and civilizations come to be based on myths.

Walter Benjamin emphasizes in particular that, in a technical age, it is the qualities of enigma and depth, which lie at the heart of memorable stories, that are threatened. Stories cease to disturb in the white light of information.

They cease to work actively in the mind when all ambivalence and richness is gone, everything obligingly "explained." To explain is all too often to "explain away." This ambivalence or quality of enigma is clearly a crucial factor in *Hamlet*, as it is in the parable of the prodigal son. It is a factor in good films as it is in good poems and good novels, and it should certainly be a factor in good television. Benjamin reminds us of the story in Herodotus of the king of Egypt who was made captive by the king of Persia. Egypt stood there and saw pass before him first his son in chains, then his daughter in bondage; and his composure did not break. Then he saw, almost at the end, among other prisoners, an old servant of his. At this sight he broke down and wept and was not to be consoled.

The story is given without further information. But why did the king mourn only when he saw his servant, and not when he saw his own children in irons? The account remains in the mind, putting its own questions, haunting the imagination. The story depends upon its inwardness with you and me, and our attempts to explore the inwardness of the characters in it.

If the central creative thrust with its inwardness and its possibilities is the important thing, is what has to be nourished and protected, then to that end the program makers need all the freedom they can get. In any case, they have to struggle against obstacles whoever and wherever they are. Like everybody else, they have to struggle against their own frailty. They have to serve their subject or theme or narrative, cutting no corners, being true to it, and seeing by taking thought as free-ranging as they possibly can that it is true to itself. At the BBC, they have a duty to the BBC and to the fact that it is the BBC's daily fought-for independence, its resources, good name, and broad shoulders they are using. A program maker, to take his difficulties still further, has to serve the audience, and that too can be a struggle and a rewarding one. But, above all, he has to struggle with the mysterious laws of creativity itself. Given freedom, he must construct his own prison; given liberty, he must kick against the iron walls he builds around himself. There never was a faster prison than a sonnet. At the same time, he must look to everything for himself. The extent to which a true creative function cannot be delegated has not been sufficiently noted. There cannot be a television program without multifarious assistance on multiple levels. But if his mark is to be upon it, the creator's hand—be he one, two, or three people—must be on everything.

John Milton has a great sentence, quoted more than once in the fifty-year-old story of the BBC: "Though all the winds of doctrine were let loose to play upon the earth, so Truth be in the field, we do injuriously, by licensing and prohibiting, to misdoubt her strength." What creativity needs is not interference, still less licensing and prohibition, but conditions in which it can thrive. What creators need are organizations or principles of action rooted in the search for, and the enjoyment of, truth, which can provide them with the freedom which is essential to them if they are to flourish and burgeon. Let there by all means be collaboration as well, but not too much.

COLLECTIVE CONSCIOUSNESS
IN MAKING ENVIRONMENT

MOSHE SAFDIE *Israeli-born Canadian architect Moshe Safdie drew international attention for his design of Habitat, a feature exhibition of the 1967 Canadian Expo at Montreal. A radical approach to apartment living, Habitat was constructed of modular housing units stacked and pivoted like a pile of blocks, to create, even at high density, diverse views and private spaces for its residents.*

Safdie's recent architectural and planning projects include, in Israel, the Yeshivat Porat Josef, a Sephardic Rabbinical College; Mamilah, a commercial/residential center; the Western Wall Precinct; the Youth Wing of the Rockefeller Museum in Jerusalem; and Jerusalem's Bronfman Amphitheatre. His projects in the United States include: Coldspring New Town, a community for 12,500 people initiated by the city of Baltimore; and urban renewal projects in downtown Stamford, Connecticut.

Author of the widely read and highly acclaimed BEYOND HABITAT *(M.I.T. Press, 1970), Safdie is also author of* FOR EVERYONE A GARDEN *(M.I.T. Press, 1974).*

CARL JUNG tells us that there is a collective sub-conscious; the sum total of all our consciouses has an identity, a power, and a being of its own. In that respect, the environment we make and live in is a mirror of ourselves anl our culture, a real exposé of what and where we are. I use the word environment advisedly, rather than buildings or architecture, inasmuch as I believe that it is the total environment that is of interest to us and not the individual building. It is the total environment and not the unique oasis that affects the quality and style of our lives. But, unlike scientists or industrialists who can look back with some pride of achievement in discussing the subjects of creativity and collaboration in their fields, the departure point in examining the subject of environment is rather discouraging.

We live in cities, towns, and villages which seem to be at the low point in man's building history. We have created serious imbalances, we have polluted the air and the waters, we have threatened plant life surrounding and within our cities, we have created cities which lack unity, which are congested, which have degenerating and disintegrating neighborhoods, which have a high level of social stratification, and which are socially incohesive. We build new housing projects which are depressive and oppressive, which lack scale and humanity, and alternately we build anonymous suburbs which speak for themselves. Our cities lack meeting places and public spaces worthy of our age.

They seem to have none of the reasons, logic, or feeling of well-being and beauty which past cities and villages had. None of us will venture to say that the man-made environment of this century uplifts the spirit of man.

I propose to examine the subject of creativity and collaboration in that context; that is, of the total environment. On the surface, the problem is simple. Even if we agreed among ourselves to skip the deliberations as to the true nature of creativity and use the word in its popularly accepted meaning and point at architects and designers with whose work we are familiar and say, "Yes, this is creative," we would certainly have to double back and make a reassessment if the question was "Is the creativity of this individual contributing toward the *total environment*, not the individual building that he has constructed?"

There are many strata of collaboration in the architectural profession. There is the architect's office, made up of people of similar disciplines though differing specializations, normally directed by an individual and working together in evolving plans for construction. Relationships in this hierarchy are not unlike those in other fields. The so-called director or chief man sometimes becomes two or three men. This happens on the rare occasion where two or three or more individuals have such an intimate, well-balanced relationship and such similar tastes that they come to function almost as one, often more effectively than one individual could. Such collaboration, not unlike marriage, is sensitive, fragile, and rewarding when it works.

Then there is the collaboration of the building team: the engineers—structural and mechanical, air-conditioning—acoustic specialists, landscape architects, contractors, technical experts, material experts, cost estimators, all a part of the building team. The collaboration among them is mandatory to the successful design of the complex contemporary man-made environment. It is clear that real creativity in the environment is not possible without crossing the boundaries to include the creativity of all these members of the building team. It is inconceivable for me to think of the realization of Habitat without recognizing the creative contribution of the structural and mechanical engineers, of the contractors who thought out the construction processes, of the technicians who solved some of the essential details. These collaborations, however, are also familiar, they exist in many other fields, be it industry or film making. They are generally coordinated and directed by an individual who has a total overview of the objectives, be it the architect, the film director, or the director of a space program.

1. Habitat '67, Montreal, Canada.

2. Habitat '67, Montreal, Canada.

3. Habitat Israel. Project for the Ministry of Housing.

There is another level of collaboration which is less
frequently considered, and that is the collaboration between the designer and the decision maker, whether the
conventional architect's client who commissions the design
of a house, or more relevantly, the group of people or
officials who are involved in the making of a segment of a
total environment, a whole community. There is a tendency to think of this relationship as a dialogue rather than
as a collaboration. There is a tendency to credit or blame
the quality of a product of the environment on the designer, but rarely do we give full marks or blame to the
decision maker himself. Many decades ago, Sullivan said,
"A building is as good as its client." An architect cannot
transcend his client. He also said, "Clients get what they
deserve." When you extend this from the individual build-

4. Habitat Puerto Rico.

5. Habitat Puerto Rico. Detail showing roof terrace.

ing to the total environment, there are interesting repercussions. Last year, we were commissioned by Sam Lefrak, the New York developer, to prepare preliminary sketches for Battery Park City, the development in Lower Manhattan. It was at our second meeting, when a discussion ensued with respect to the basic principles in question, that Sam turned to me and said, "Moshe, you must know the golden rule, 'He who has the gold, makes the rules.'"

For the past two years, my office and I have been intensely involved in the development of the new town intown of Coldspring, in Baltimore. We are about to start the final planning work leading us to construction in six months, the developer having finalized his agreement with the city. I believe that Coldspring will break new ground in the creation of a community constructed within an existing metropolitan city. But, in retrospect, examining the activities of the past two years, I can say that the uniqueness of the experience and the results are a product of the collaboration which took place between the design team, represented by the architect/planner; the municipal government, represented by the housing commissioner, Robert Embry, who, in turn, represented the community, which was also actively involved; and the developer, who represented the constraints and realities of the market place, of industry, and of management.

Coldspring will be possible because, as a by-product of this collaboration, creative solutions to the problems of financing, city administration, architectural design, community participation, and community management have all come together. Without any of these ingredients, Coldspring would not be possible. It took this kind of collaboration to solve the matrix of problems in breaking new ground to create this new town. If there is a uniqueness to this collaboration and its product, I would attribute it to

6. Coldspring, Baltimore, Maryland (a new in-town community for 12,500 people).

7. Coldspring. Clustered housing designed for steep-slope conditions.

the fact that the city government, the community, the developer, and the architect and his team had common, clearly defined objectives, and that all those involved could get together, catalyzed by the initiative of the city government and its stated objectives. These objectives had to do with dissatisfaction with the normal commonplace solutions to building a community in the context of the urban environment at the physical level and the social level, and the desire to seek alternative physical, social, and economic structures.

Let me be specific. It was recognized that within the present economic structure, given the market cost of land, money (interest rates), and construction, the earning and related purchasing power of the majority of Baltimore families, including a wide segment of middle-income families, could not afford well-designed new housing.

—It was believed that piecemeal development by many developers results in social and physical chaos.

—It was stated that the prevailing standards of dwelling design of middle- to higher-density housing was inadequate for family living.

—It demanded positive solutions to such problems as the car and its storage, separation of pedestrian and vehicular circulation, a usable park network, and the provision of adequate shopping and community facilities.

—It recognized that the quality of schools, health services, and community life were as important as good physical design.

—It proposed that there would be a mixture of families of varied income and race in the community, housed in dwellings of varied types and densities.

—It assumed that only active community involvement in the design process would create a strong identity and, later, a sense of community structure.

True, the developer also had other objectives, that is, to bring about a return on capital invested, and city officials might reasonably have the objective of re-election. Various community groups represented in the design process even had conflicting objectives, but all were united by a strong conviction, an overriding objective for making an improved environment, the characteristics of which were well defined.

It is the uniqueness of this situation, however, that should disturb us. There will always be the rare enlightened government official who will use the bureaucracy as a tool in implementing the needs of his constituency rather than as a shield and an escape. There will always be the enlightened developer who will seek creative ways to combine a profitable business and building that which is inspiring to those who live in it. And there will always be archi-

8. San Francisco State College Student Union.

tects whose articulation of objectives is such that they will be able to act as equal partners in this team. But as these situations are rare, the total environment remains the mess that it is today with only a few exceptions. So if we find the breakthrough in making environment requires collaboration of the highest order, we also find that the lack of clear objectives is detrimental to the existence of this collaboration.

James Watson made it clear in his lecture that, at any given time, the next problem in science that has to be cracked is known and clear to everyone in the field. Akio Morita also emphasized that, in industry, it is always obvious where inventiveness and creativity will be needed to discover new solutions. What remains to be done is to solve these problems.

But, in contrast, in architecture there is never agreement

9. San Francisco State College Student Union. Detail of exterior.

on objectives. Let me be specific: Architectural competitions call for many architects to come forth with solutions to a defined environmental problem. Even when the programs of such competitions are well defined, we often find (alas, even expect) contrasting and contradictory solutions coming forth from different architects. Even more disturbing, consider the problem of the design of a university campus or a new town. Consider the possibility that an enlightened authority should select what they consider to be the world's greatest five or six living architects to work on the design of such a complex. It is certain that these five architects, the creativity of whom we would not dare question, could never find the way of working together to create this unified and whole environment. On the contrary, the more "creative" they are, the less likely they are to be able to collaborate to find common objectives and a common language. Consider the work of the greatest architects of this century and try to put them together in a single environment. It is not only the issue of a contradiction of styles, but it is the fact that there are no agreed-upon common denominators, there are no agreed-upon objectives which override the idiosyncrasies and pecularities of these so-called creative individuals. There is not even a common cultural bond to serve them in their task.

So while we see in other disciplines that collaboration is possible once the individual psychological and ego-related problems are resolved, we see that in the field of the environment, collaboration seems almost impossible and, at face value, *contradictory to the commonly accepted definition of creativity.*

Consider two situations: An Arab peasant living in his village somewhere on the hills by the Mediterranean decides to build a house for himself. With the aid of some of the village craftsmen, he sets out to select a site and construct his house utilizing the vernacular building traditions known to him. Some miles away, in the modern metropolis, a university professor selects the most noted and reputable architect and together they set out to design and build the professor's residence. Though the peasant has never been professionally trained as the architect has, and the chances are that he never even attended school and may not even be able to read or write, it is quite probable that the quality of the environment of the peasant's house will be superior to that of the architect's and his professor.

Without any doubt, we could state that there is no question that, when considering the total environment of the village, the peasant's house would relate better, and form a unity and a whole which would contribute to the total environment in a way superior to that of the architect's house to its surroundings. And so, who is more creative and how do we measure this creativity? Is the highly trained architect more creative than the peasant? What validity do these terms have when we consider the total environment?

This dichotomy has forced me to backtrack and examine the history of building in an attempt to understand the relationship between creativity, collaboration, and the environment. There are some basic characteristics of the work of our peasant and our architect which start to give us an inkling of the problem. The work of an architect tends to be characterized by a highly arbitrary design process. Because of its arbitrariness, we find that style and fashion play an important role in design. But fashion is arbitrary; it puts a time limit on beauty. Arbitrariness of style and fashion is contradictory to the design which has a true fitness of purpose, and, by necessity, makes it impossible for the part to become an organic unit of the whole. In contrast, our peasant's work is a by-product of a vernacular building system which has evolved over centuries in response to the

properties and limitations of available materials, climate, topography, and life-style. It therefore has a nature-like mor-
phological fitness of form to needs. It is not arbitrary, its
style and culture are the by-product of a response to need.
Both the peasant and the architect represent historical approaches to the design, conception, and construction of the
environment.

In the tradition of our architect, architecture is thought
of as an art form. A word of top priority is expression. Our
architect's predecessors designed and built for the power
structure—temples, tombs, palaces. Key words in their lex-
icons are composition and proportion.

Our peasant's key word is building tradition; we might
use the word vernacular. The basic characteristics of this
building tradition are evolution and integrity of materials—
responsive to climate, to life-style, bound by rather static
cultural patterns wholly accepted by the peasant and his
society.

In both traditions, we find the search for common denominators. In the peasant's tradition, the search is not a
conscious one; rather, it is a by-product of the process.
There is no question that we recognize, and are aware of,
the common denominator that binds together and gives
unity and wholeness to the Greek island villages, to the
Italian hill towns, to the Indian pueblo, the village architecture of the Chinese plains, or the New England villages
and towns. It is a unity based on a vocabulary of form
which is almost as unquestionable as the forms and organisms of plant life and animal life.

But, in our architect's tradition, there has also been, in
the past at least, the search for common denominators. To
be sure, they are basically stylistic. They usually rest upon
the complete acceptance of certain formal systems, I would
rather say formalistic systems, be they classical orders with

all the rules accompanying their use or, in more recent examples, the massive façades of the architecture of the Third Reich or the Mies van der Rohe steel-and-glass curtain-wall towers of the 1950s and '60s. But there always appears to be, even on superficial examination, a high price to be paid for formalism. Generally, these formal systems are imposed at the cost of responsiveness to the very elements that created the common denominator of the architecture of the peasant. Steel-and-glass office buildings ignore the asymmetry of the sun's movement in the sky, the difference between north and south on the face of a building. The need for an individual's or a family's identity in the total complex of a community is sacrificed on the altar of the formalistic system.

The history of the modern movement in architecture of this century reflects a conflict between these two building traditions. In the twenties, we have the radical statements, the Bauhaus, the rejection of formalism of any kind, the concentration on process, hence the industrialization of building and the preoccupation with the social problems of planning cities. And yet, in contrast, the practitioners of the modern movement were involved in building structures, not total environments, which were very much in the tradition of the artist-architect, very much composition in three dimensions, whereas the philosophy set forth was that form followed function. The word aesthetics had high priority in their lexicon, and while they talked about the industrialization of building, their buildings in fact were built by handicraft methods. In the forties and the fifties, the formalism we are now familiar with of Miesian steel-and-glass towers, of so-called universal spaces, began to emerge. But the universal space, designed to give ultimate flexibility to its user, anonymity to the designer, and determinability to the user, in fact were the least flexible of all. Note, for example, the severe restrictions on the use of curtains and other kinds of paraphernalia which have always been imposed on the users of those buildings. And then came a reaction, deeply rooted in the dissatisfaction with the inhumanity of the new formalism. But, again, this reaction was mainly based on fashion and style. It talked of an architecture of delight, a gay architecture, but, with attempts to break away from the constraints of the formalistic precedent, came a wave of arbitrarily fashionable designs reminiscent of automobile grilles, which bring us back to our total absence of a common denominator.

And it is here that I must come to what I feel is one of the basic issues surrounding the contemporary environment. In science and industry, the objectives are clear and therefore collaboration is possible. This collaboration is not only among a small team of individuals working together, but it is a *universal* collaboration of all the many minds, many teams working toward similar goals. But, in architecture, it is irrelevant to talk of the collaboration within the small building team itself. The real issue is that there is no *universal* collaboration of all those working toward similar goals in which the efforts of each individual, each team, each group are cumulative and additive toward these universal goals. This universal collaboration, or, rather, the lack of it, is what stands in the way of creating an environment better than that we have today and perhaps as good as that we had yesterday.

My interest in the peasant tradition is not merely due to a desire for visual unity in the environment; there is more at stake. The apparent success of the peasant to act creatively raises critical questions for us: In every field and in every discipline, there are more- and less-gifted people. Science and medicine have their geniuses and their hacks and so does the architectural profession. And since, statistically, the number of mediocre practitioners will always greatly exceed that of the uniquely gifted ones, we face in the environment a problem not encountered in other disciplines. Mediocre scientists are not detrimental to the progress of science, but since the majority of buildings are destined to be designed by mediocre designers, we must make the obvious supposition: We are doomed to live in a mediocre environment which, from time to time, is highlighted by the rare work of a gifted and creative designer. If we accept the present methods of functioning of the architectural profession, we must also accept this supposition. In this book of definitions, in which architecture is art and creativity is artistic, the opposite is also true: the mediocre is mediocre and the chances for an improvement of the total environment are nil.

So, perhaps, by exploring the means by which we can bring about a universal collaboration in the process of making the environment, we shall also bring about a redefinition of the nature of creativity.

And in doing so, we might have to make certain adjustments in the whole cultural perspective we have of the place art has in our lives, and in its relationship to non-arbitrary scientific endeavors.

Rutherford was once asked how long he believed it would have taken for his discoveries on the structure of the atom to occur had he not existed. "It would have inevitably happened within a few years," he stated. While most of us would consider this answer reasonable, we would nevertheless be inclined to assume that in the absence of a Bach or a Beethoven, a Shakespeare, a Dante or a Picasso, the history of music, painting, and literature would have fundamentally changed, and so it is in architecture. We assume that in the arts (we include architecture), the role of the individual's creativity is different from in the sciences.

I would like to submit that this is not the case. It is possible to separate the basic structural qualities in the work of the masters in the arts which are inevitable sequences in the evolution of the history of culture, from the particular personal envelope, which is of secondary importance. Hence, the course of music, literature, and, most relevant to this discussion, the history of building are inevitable evolutionary processes and the role of the individual is no different from in the sciences.

Architecture thought of in the context of art has been the curse of the profession. Recently, Yoram Kaniuk, the Israeli writer, remarked, "It is the difference between the architect thinking of himself as the *master* of his buildings or the *servant* of his buildings." It is the master tradition that has brought about the distortions that exist and it is only the servant attitude that would put them back into their proper perspective.

The division of art and science in Western culture has been the subject of much discussion recently. To relate it more specifically to environment, let us examine the relationship between building and the visual arts. Our peasant's house and life in his Mediterranean village is rich in art, or shall we call it artifacts: his carpets, rugs, textiles, clothing, saddles, jewelry, pottery, metalware, and the objects of religious worship enrich him and his environment. But he never thinks of these as art, nor does he call their makers artists. These objects enrich environment, but they have two basic qualities: they are objects necessary for daily life, and they are part of the physical system of the environment.

Our professor's house, particularly if he has the financial means, is filled with art: painting and sculpture. More relevantly, our urban plazas, lobbies of corporate headquarters and hotels are almost conceived of as a setting for art. It is felt that the environment of a room, a house, a plaza, is not rich enough, it must be embellished. In turn, our objects of daily life, cars, fabrics, utensils, for the majority of people, are not looked upon as a source of art in their life. This artificial separation between art and environment directly relates to the possibility of collaboration in making environment and to the role of the individual versus collective creativity. If you consider a total and unarbitrarily conceived environment such as the spaceship, it is neither possible to think of paintings hanging on its walls to embellish it, nor the need for them, any more than it is possible to conceive of separating the activities of its functional design and aesthetic design. It is also clear that it is easier for us to comprehend collaboration in the conception of a spaceship than in a house or a building, but the same laws and processes apply to both.

The first step is to eliminate the arbitrariness in the design process. Instead of thinking of architectural design as expression, we must think of architectural design as problem solving. This involves a complete reorientation of the profession, the training of architects, the self-image that they have, and more importantly, the image that society has of them. It is society that asks the architect, "What have you done about aesthetics?" "What will be the cost of aesthetics in this design?" Or "Your design may be functional but is it beautiful?" It is society that accepts the illogicality of the possibility of something which is functional but not beautiful. It is society that uses the word functional in such a limited sense as not to include in it, or extract from it, the whole spectrum of problems relating to the psychic well-being of man in his environment. That is, his need for privacy and identity, his perception of space, color, and texture, his feeling toward scale, and his need for a sense of spatial hierarchy. And so, by having the word "function" include all these, by insisting that the design must grow in response to program, and program become a detailed, precise, and well-thought-out statement of needs at the physical, psychological, material, structural, and every other level, we can avoid the mess we live in. The more precise the program, the less of a chance the architect has to be arbitrary, the less will be his reliance on formalism and other preconceptions. The more he responds to real needs, the more likely his architecture is to have common denominators with that of others dealing with the environment.

I do not suggest here that we are about to evolve such complete programs, that the design process will become an automatic one, that building processes will be so well articulated as to lend themselves to being fed into a computer and, in logical sequence, produce building designs that are unquestionable. The environment is far too complex, and our knowledge and understanding of it far too limited, to bring about any such precise definition that would guarantee results.

The second method by which we could bring about universal collaboration is the creation of environmental standards. I first became aware of the need for standards working here in Washington on some of the earlier concepts for the Fort Lincoln New Town. We were several architects working on separate design solutions to be built within one demonstration project. The man in HUD responsible for the program was, perhaps fortunately, not an architect but a physicist, Dr. Tom Rogers, whose previous background was in defense research. He put the question to us, "When your proposals are complete, I will have to evaluate and measure your designs in order to make recommendations for the future; how am I to measure them?" It was through this question from a member of the team acting in a true spirit of collaboration that I came to understand that his inability to evaluate our proposals easily was the result of the absence of any clearly defined standards of environment. Had such standards existed, we would have had to respond to them in our design and he would have been able to measure our design by its success in responding to the environmental standards. This conversation prompted me to prepare the first draft of the environmental standards for the design of a residential community. Of necessity, I had to put down on paper those characteristics and qualities of environment which I felt were mandatory for the success of this community. They ranged from such matters as the acoustical and visual privacy between neighboring dwellings; the nature, size, and quality of the access space to dwellings; the arrangement of the dwelling itself; the community facilities that must be incorporated; the need for outdoor, and the nature of indoor, space. Given the problem of a residential community, I soon accumulated hundreds of environmental standards. As in the case of a building program, I do not feel that standards guarantee the solution. Standards can be dangerous and stifling if they are stated statically as solutions rather than as goals. But environmental standards, which I later came to rename "Environmental Code" and once again renamed "Environmental Bill of Rights," thus giving them a political context, are the only hope we have to establish a setting in which a number of individuals, who are part of the design and decision-making process, can have a dialogue, make decisions, and collaborate. It eliminates the mystique that the architectural profession has wrapped itself in, a mystique that states that the measurement of the success of a particular design is a difficult and subjective process, that architecture, like art, cannot be measured precisely. This mystique, while protecting the architects, has really been to

10. Yeshivat Porat Joseph, Jerusalem, Israel (rabbinical college).

their detriment. It has weakened the architect in any dialogue or conflict as to what should or should not be done in a particular design problem. It has put him on the defensive in the never-ending, inevitable, if healthy, conflict between *environmental objectives* and *economic realities*. In the absence of environmental standards, the architect's desire to uplift the quality of the environment, often involving complexity, is dismissed as a whim—the architect's idiosyncrasies, his funny aesthetic ideas. His bureaucratic counterpart, always striving for simpler and therefore cheaper solutions, can compromise what is not a whim but a fundamental aspect of environment clearly measurable by the environmental standards.

The third area in which re-examination might bring us toward universal collaboration in making the environment

11. Yeshivat Porat Joseph. Interior of the synagogue.

is the tool of *the master plan*. Master plans are an ancient tool. When used for small-scale projects which are built over a short span of time, they come to resemble building plans. But when used to plan large environments built over a long span of time, they have come to possess characteristics of their own. The traditional master plan has nothing to say about the quality of environment. To generalize, take a land-use plan. A basic road system is delineated, and large patches of color—yellow for housing, red for commercial, purple for industry, green for open space—indicate some sense of quantity but no sense of quality as to what is to be constructed in this town, neighborhood, or city. I became acutely aware of that fact when serving on the Jerusalem Subcommittee of Planning, which in 1970 was asked to review the master plans prepared for the Jerusalem municipality to guide the city's growth in the coming decades. The presentation of the plans was preceded by verbal statements with which no member of the committee could quarrel: the need to keep the city compact, to respect the present dividing line between the urbanized and the rural, to respect the identities of the mosaic of neighborhoods that make up Jerusalem, to build with respect for the topography on a scale which relates to the scale of historical Jerusalem. But the plans that were presented did not say these things at all. Underlying the structure was a road system that, at best, was totally unsuitable for an existing historic city with a sensitive and vulnerable urban fabric. Complementing this was a traditional land-use demarcation. But the yellow of the housing told you nothing of the quality of the communities that would be built on the lands painted yellow. The red of the commercial told you nothing of the character and quality of the public meeting places and commercial areas. Would they be marketlike bazaars? Would they be massive shopping centers sur-

12. Western Wall Precinct, Jerusalem, Israel.

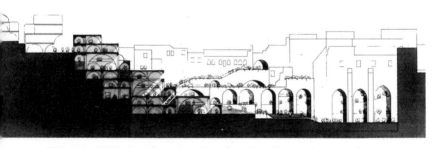

13. Western Wall Precinct. Section showing the vertical and horizontal progression of scale.

rounded by acres of parking? Certainly, that which was under construction at the time the committee met, and which was therefore visible, created great concern.

In my own mind, the argument was not with the plan,

14. Western Wall Precinct. Axonometric view.

228

the argument was with that which was not in the plan. The plan made quantitative statements, some questionable, some not. But it did not attempt to make any qualitative statements. I believe that master plans can be changed to become more-comprehensive, qualitative statements which will also be the basis for collaboration between all factions, all interests, professional and political, in determining the fate of the city. There is no question that there are certain elements that have powerful structuring effects on the environment as a whole. These are: the transportation and road networks, the location of major transportation facilities—rail lines, airports, harbors, arterial road systems, and all other means of public transportation. Important, too, is the broad separation and determination of what land is to be urbanized and what land is to remain as public open space or agricultural. The transportation network and the demarcation of open and urbanized land set the tone of the pattern of settlement, the size of communities, and the density of these communities, but as we come to the development of each neighborhood, we must inject new characteristics into the master-plan documents.

The city of Baltimore recognized this in establishing the terms of reference of the design of Coldspring. It stated that the traditional land-use plan is a useless master planning tool for a neighborhood of this size. Building design, at its broadest conceptual level, and master planning cannot be separated, but, on the contrary, must be integrated into a single process and document. It is building patterns that generate the road system. The location of buildings—the decision as to whether to build on the flat lands or the slopes—the determination of the size and nature of open spaces, must establish the height and mass of buildings, the circulation networks, and the system of access.

In developing the master plan for Mamilah, twenty-five

15. Mamilah, Jerusalem, Israel (a new commercial, entertainment, and residential center, linking the Jaffa Gate and the New City of Jerusalem).

acres in the heart of Jerusalem adjacent to the Old City, we went even further. In addition to establishing a three-dimensional network of land uses, a precise definition of massing and height, a complete system of access and circulation, we proposed a vocabulary of building elements available to any architect building within that district, working for any developer who might be designated in the future to build within the context of this expanded master plan. The master plan thus deals not only with scale but with program and its subtle requirements and determines the qualitative results that will be generated by the developments. I believe that I am only scratching the surface of what this development might mean in the future. It cer-

16. Mamilah. View from the south showing the Valley Park.

tainly means that much of the decision making in the environment which is always made after the fact—decisions at the municipal level, at the developer level, and at the architect level, which are made as a reaction to a situation—will be made beforehand when a total and comprehensive review is undertaken at the master-planning phase. This three-dimensional master plan—urban design and qualitative program—must be taken beyond the sketchy outline I have stated here and applied neighborhood by neighborhood to existing neighborhoods, to neighborhoods in change, and to new neighborhoods which are yet to be built. It must also fit into the greater total structure of the regional plan.

Much of what I have said seems to suggest the breakup of creativity into a hierarchy: the differentiation between creativity at various phases of making the environment. Rather than consider the romantic notion of the creative designer, we now have a series of overlapping interpenetrating creative acts: the creative act of making a program for building an environment, and the responsive creative act of synthesizing that into building form; the creative act of establishing environmental standards, and the creative act of responding to that with building form. One kind of creativity is involved in making the comprehensive master plan, and the responsive creativity is in interpreting that into specific building form by one or many architects. And it is here that I would like to come back to our peasant. He, too, related to the design process in such a way, except that the building vocabulary which he inherited was not developed by an individual and was not the result of individual creativity, but, rather, was the result of collective creativity evolving over a long period of time. It is that collective creativity which created the building vocabulary of domes and arches and courts which responded well to the needs of this peasant's environment. It is this collective creativity acting in cognizance of individual creativity that we must re-establish in our own cultural tradition.

We have a highly fashionable and popular contemporary term that leads us to the possibility of establishing this process in our own activities, that is, *building systems*. Though the term was coined to deal with technical systems of building, that is, building processes, it has now expanded to mean much more than that. Building systems are, in fact, environmental design systems. They are no different from the peasant's best building vocabulary. That, too, is a building system, both in the technical and the environmental sense. And so we might find that we should devote our creative energies toward the evolution of building systems, responding to the multitude of building needs: communities, schools, hospitals, urban centers. We should devote our creative energies first to clarifying the program for such buildings, then synthesizing it into a vocabulary of building form. Key words, evolution and additive information, are characteristic of this process. At present, every time an architect designs a school or hospital, he must reinvent the wheel; every time he does so, he has no access to the inventiveness and experience of all those who preceded him along this path. In evolving these systems, each of us will add to the work of those who preceded him. And then, on a mass basis, wherever the need might be, these building systems can be applied by individuals, by groups, professionally trained or not, responding to their specific needs. The closer the client is to being the user himself, the more precise would be the definition of need and the more likely would be the response of the design to it. The more distant the client from the user, the more bureaucratic, the more abstract, the less likely he is to understand the needs of those who will use it.

If I have left parts of this proposition vague or unclear, it is because it is also vague and unclear in my mind. I do believe that the untrained peasant who built his own house has something to tell us. I do believe that there is the possibility of evolving our own contemporary vernacular. I do believe that it will give our cities the sense of unity, a sense of response to need, a sense of truth beyond what we have today. It would be the result of searching for the common objectives and would give us the common denominators which we lack. But, in closing, it is also necessary to speak of means, of energy; one cannot avoid doing so, speaking in Washington. Because I believe that, as we clarify objectives, we will also discover that we are lacking the means, in terms of resources, in terms of money, to meet and respond to these needs. There are no magic solutions, there are no miracle technologies, there are no secret materials which are going to be discovered to resolve the gap between our desires for the environment and the available means to meet them, between the definition of what is a good house and the available funds to build it within the earning power of a family or a community. Eliminating pollution, cleaning up industry, building spacious houses, giving a garden to everyone, creating inspiring urban meeting places, penetrating the city with open park space, all cost money.

But here, too, there is reason to be optimistic. As long as these objectives are vague, it is easy to allocate resources to other matters which appear to have priority. But as we come to clarify them, to clarify objectives, and as these become not those of an individual in the design team, the architect alone, but those of all the collaborators on that team, of society itself, they also become more difficult to resist, and then, as we have the will, we also are likely to find the way.

CREATIVITY AND COLLABORATION IN GOVERNMENT—THE BUDGET PROCESS

CASPAR W. WEINBERGER *Since coming to Washington in 1970 to head the Federal Trade Commission, Secretary of Health, Education, and Welfare Caspar W. Weinberger has held a succession of top-level posts in the Nixon and Ford administrations. Less than a year after taking over the FTC, he shifted, again at President Nixon's request, to the newly formed Office of Management and Budget as Deputy Director. In June of 1972, he rose to the directorship of OMB. He was appointed Secretary of Health, Education, and Welfare in February of 1973.*

Before coming to Washington, the Secretary had had a varied career in his home state of California, as attorney, writer, television moderator, state legislator, and State Director of Finance.

Now that we have considered creativity and collaboration in industry, bio-medical sciences, in television, architecture and our environment, we come, at the end of this series, to discuss creativity in government.

The terms may seem incompatible—or mutually exclusive, even antithetical.

Is not creativity either undesirable in government—or, depending on the viewpoint, stifled and blocked by government? Indeed, Huw Wheldon, in his brilliant lecture in this series, said that "Independence from government is crucial if creativity is to exist at all."

But he was speaking of the BBC, and tonight we want to consider creativity *in* government.

Classically and traditionally, one should start with definitions—a beginning that has thus far been avoided by my predecessors in this series and probably for good reason, given the elusive problem of trying to define creativity. Mr. Wheldon, of BBC, said he "flinched" from words like "creativity." Moshe Safdie spoke of using the word creativity "in its popularly accepted meaning." Mr. Morita, of the Sony Corporation, left it simply by saying, "We all know creativity is vital to the progress of mankind." And so it is, and that states one of the cardinal tasks of government: "We have to work out a society that is sufficiently stable and rigid to support the human being and allow him to function out of its rigidity, but at the same time sufficiently elastic to permit him to go beyond its bounds."[1] Stated more simply, the central task of government is to create a society that combines freedom and order, and *that* calls for a high degree of creativity indeed.

Many of our governmental institutions sprang from the creative impulse of the founders of the republic. The combination of the House and Senate, with their differing bases of representation (a structure our courts will no longer permit our states to use), the Supreme Court itself, the presidency, the separation but interrelationship of the three branches, our federal system—all of those were born of creative skills, and evolved out of our experience and the stability our government fostered.

Creative people are concerned with the ideal, with the abstract world of ideas, philosophies, new concepts—with designs. It is "inwardness which characterizes an original work," as Mr. Wheldon says.

The creative person does not worry about obstacles— nor raise them. The creative person searches for solutions, for such a person never believes problems, no matter how difficult, cannot be solved.

Creative individuals see more complexity in things that concern them; they are open to greater perception. As Jung puts it, the creative person perceives more, becomes aware of more than non-creatives, who make judgments about their material within reasonably narrow confines. The creative person has virtually limitless vision, because he has limitless horizons.

Part of this breadth of vision is achieved by reliance on intuition—a characteristic of the creative person.

But so little is known about all this that while no one can speak with any certainty about creativity, it is clear that it is the next great frontier of the mind to be explored. Indeed, this expedition into the realms of inner space has already begun.

Several studies of the right hemisphere of the brain—

thought to be the seat of the artistic, creative, non-verbal talents of the person—are now underway, after several years' study of the left side, home of verbal, analytic, cognitive-learning skills. The left side of the brain enables us to measure or analyze, but the right side is where we now believe *thought* originates.[2]

And Dr. Paul MacLean's studies of the new mammalian brain may give us even more clues.[3]

Incidentally, many of these studies themselves are excellent examples of the creativity that is *fostered* by proper governmental encouragement, not domination.

Meanwhile, until we learn much more, we have to content ourselves with observing and identifying creative ideas in government, and in all other manifestations, without quite knowing where they come from.

Until we know, our task must be to foster creativity in the only way that we know, and that is by bringing together the most innovative, complex, intuitive people we can find and giving them maximum freedom and encouragement to think, to exchange and test ideas with competing ideas in the crucible of healthy skepticism.

Even creative endeavors can be fully organized, or, to phrase it another way, creativity can be joined with the collaborative process. Winston Churchill, who had one of the most creative minds in modern times, wrote his great biographies and histories with the aid of many, whom he organized and whose product he inspired and then quite properly used. Their work, fostered by his organizational talents, enabled him to create his works far more quickly and effectively.

In science, too, as in so many other aspects of modern civilization—industry and art, for example, as we have seen in this series—creativity is now organized and combined with the collaborative process.

Large pools of talent, facilities, and resources are brought together—frequently by people whose skills are considered non-creative but who may have a talent for accumulation or management.

We see this also in the accumulation of many skilled talents assembled by government to reach a common goal—for example, the massive effort now underway to advance us toward the ultimate cure for cancer in all its hundred forms.

There is another area of government where creativity may not be as apparent but where both creativity and collaboration are an essential part of the process.

It may seem a long way from creativity to the budget—but in the federal government and in other governmental levels, the most basic decisions that affect policy in all executive departments, and indeed in the entire economy for years to come, are made in the budget.

For many people, in and out of government, the budget is a mysterious, little-understood phenomenon. It nearly always makes people unhappy, for there is never enough for those who believe spending is the solution to all ills; and there is always too much spending for non-Keynesians.

But while it is not possible to please everyone, or indeed usually anyone, with the budget, it is possible for far more people to understand the budget process and, thereby, to see creativity and collaboration in government in action.

The budget is a number of different things: In state and local government it is a collection, usually in one bill, of proposed spending for existing programs together with an estimate of the revenues that will be received from the existing tax structure. In the federal government it is this and more, because the budget of the federal government is so large, being over 20 per cent of the gross national product, that it has a major influence upon the state of the economy for any given year in which it is proposed.

Actually, I believe that the influence of the federal budget upon the economy is greatly exaggerated, but you must be aware that this is distinctly a minority, or non-Keynesian, view. In any event, the budget is regarded, usually by people who have nothing to do with making it up, as being a means of influencing the national economy.[4]

These people tend to think of it as an instrument that can be finely and accurately tuned so that if there appears to be a recession or worse in the offing—or upon us—the virtually universal remedy is to increase spending. This, it is thought, will stimulate the economy and ward off the recession or worse. By the same token, and by some, but not many, of the same people, it is felt that if the economy is already booming along and there are evidences of inflation ahead, then we should restrict spending, and all will be well.

These views are so firmly and widely held by so many economists that any challenge to them is usually shouted down. Nevertheless, and at the risk of incurring such evident displeasure, a few words as to the fallacies of this theory, whose best-known advocate was Lord Keynes, would seem to be in order. For although this was a very creative theory by a very creative man, it demonstrates that creativity alone is not enough. All creativity is not necessarily good. Indeed I happen to believe Keynes was wrong, because I simply do not believe increasing or decreasing spending for federal programs is a very effective way of managing the national economy, either as a cure for excessive inflation, or unemployment.

We must look to other ways of solving our economic difficulties than an unthinking resort to budgetary manipulation, particularly on the spending side. If, instead, budgets were developed on the merits of programmatic proposals, not the current economic outlook, our planning and administration of programs would be more effective. But spending more public money either for the sake of spending it or to help the economy seldom if ever helps.

There are four main reasons why the basic Keynesian model is no longer adequate for today's economy.

First, the causes of unemployment and especially of inflation are no longer influenced as significantly by a budget deficit or surplus as they once were. International conditions, private collective-bargaining arrangements, and scientific discoveries, just to name three, have a far greater influence on our economy today than whether the budget has a deficit of $10 or $20 billion in an economy with the GNP approaching a trillion and a half.

For example, it certainly is not the budget policy of the past three years that has caused a tripling in the price of ferrous scrap and a doubling of wheat and cotton prices, to say nothing of the price of a gallon of gasoline. Nor is the unemployment at service stations or automobile assembly plants going to be cured by an increase in aggregate demand. Increased spending may, and I think does, fan the flames of inflation, but restraint does not put them out—if real budget restraints were ever possible.

The second difficulty in applying the Keynesian model is political. The idea of spending more federal funds—regardless of revenues—when economic activity drops, and spending less when economic activity expands and inflation threatens, is indeed a clever and beguiling theory. If you could do it, it might even work. But Lord Keynes never had to go up Capitol Hill to convince a majority of any party on any committee of either house of Congress to support budget reductions at any time, let alone when his theory called for budget reductions.

Anyone who has tried to reduce appropriations once

voted, because of changed economic conditions, or to vote a smaller budget than last year, or to terminate existing programs, knows that the "restraint half" of the Keynes model is impossible to execute. It is impossible because the collaboration of the Congress, which is required if the theory is to work, is wholly absent. And, if you proceed without the collaboration of Congress, I can testify personally you will be named defendant in innumerable law suits, charged with creating a constitutional crisis, among other things—when you try *not* to spend money that the Congress has appropriated. Indeed, if you do not request annual *increases* in amounts equal to the Japanese annual GNP growth, almost everyone discusses you in strong rhetoric, in editorials, picket signs, and similar critical journals.

Thus, it is all but impossible to reduce spending when the model calls for that. Of more than one hundred programs recommended for reduction or elimination in 1974, virtually none were eliminated and less than 10 per cent were reduced.

Furthermore, on the other side, it takes so long in our system to set increased spending in motion that the economic need for it is often past by the time it takes place. We do have some programs, such as welfare and unemployment insurance, whose outlays naturally rise and fall as the Keynesian model would want them to do, in response to changes in general economic conditions, but these do not amount to a significant portion of the budget. This is also true on the tax side. But, to increase expenditures in most programs takes a year or more.

Let's assume that a large budget is proposed in January to counter a sluggish economy. Because of the substantial lead time to get proposed budgets enacted and spending from appropriations into the economy, theory never catches up with reality. It would be far more likely, in fact, that the big budget would be counterproductive in the end. The full force of the increased spending probably would not hit until natural economic forces already had picked up the economy, so that the increased spending, when it finally does take effect, actually encourages inflation rather than building up a flagging economy.

I am reminded of that great, but I suppose mythical, Pentagon slogan: "If it works, it's obsolete." Recasting that thought to apply to the budget, it says, "The budget now in effect reflects the economic conditions of two years ago, not the realities of today."

Highway and public-works construction, for example, are often proposed—by the construction industry—as one way to "get the economy moving." Yet, most of this spending occurs no earlier than one and a half to two years after the appropriation. For in addition to the necessarily prolonged consideration before enactment of an appropriations bill in the Congress, much time is involved in developing detailed work plans and contract specifications, and the letting of contracts, with all the paperwork and seemingly endless review that entails. And finally, of course, the contracts have to be performed before the economic effect is felt—and that means years.

A still more serious difficulty with the politics of the Keynesian model develops when reductions in outlays are called for. As we have seen, these are simply impossible to obtain in a Congress beset by special lobbyist spending groups and virtually no "don't spend" lobbyists.

The fact of the matter is that the spending side of the budget cannot be turned on very fast, and it is virtually impossible to turn off.

Thirdly, the budget is simply not a precise enough instrument to "fine-tune" the economy. Far from being a Stradivarius, as common lore would have us believe, where each $100 million is identifiable and controllable, it is no secret in Washington that the budget is more like a radio whose volume knob can only be turned up, not down, or an organ whose pedals are too often unlabeled and stuck together.

It is virtually impossible to control the rate of spending in the many departments, even if there were common agreement, which there is not, as to whether signing a contract or paying for performance was the significant "spending" in terms of generating economic activity. Often, months or even years pass between these two events. I have seen estimates of outlays that have been off by as much as $4 billion with only five months to go in a fiscal year, and outlays are much easier to predict and control than revenues. The degree of collaboration needed to control even the rate of spending has not been achieved.

So, even if budget policy could operate effectively on the causes of unemployment or inflation, which it cannot, the budget as a practical matter is not manageable within narrow enough limits to be confidently used.

The fourth and probably most important reason to resist the proposals for altering spending patterns in response to economic conditions is that it generally is bad for the programs involved. I believe that we should reach decisions on program levels on the merits of the program. Planning and administration of programs are seriously hampered, if not made impossible, by a succession of tight and loose budgets. Evaluation of programs, a fledgling but essential idea in Washington, is rendered pointless.

Every year should be a tight-budget year in the sense that the taxpayers should get their money's worth. But they do not get their money's worth if we defer a space-program project that is going to cost more when we eventually do it, or if we push money out indiscriminately on the basis of results from ill-conceived sociological research. In the human-resources area particularly, undisciplined spending can be counterproductive. When a larger share of the budget was devoted to the straightforward purchases of goods, at least a loose budget meant only that we got more than we needed. In the human-resources area, we sometimes get more of what we did not need in the first place.

It seems to me, in short, that priming the federal pump—or unpriming it—simply cannot be done flexibly enough, efficiently enough, or fast enough, to make the necessary difference soon enough. Or, as someone said, "Even Lord Keynes wasn't a Keynesian when he died."

In any event, logic aside, it must be conceded that the Keynesian view that you can and should manipulate the budget to affect the economy is so widely held that we must start with it as part of the federal budget process.

We must also understand, for understanding is the beginning of wisdom, that the procedures in the Congress, although at long last undergoing a comparatively serious self-examination, which has resulted in a change, are incredibly bad. No self-respecting state could or would live with the absurdities of the Congressional budget process. To begin with, the Congress, once it receives the President's budget, chops it into fourteen pieces which are considered by at least three times as many committees in a wholly piecemeal, uncoordinated way.

Thus, despite the widely held belief of most senators and congressmen as well as economists and even taxpayers that over-all budget totals are important (and indeed essential if you are going to manipulate the federal budget without regard to the necessities of various programs, to try to alter the economy in any given year), there has been no Congressional method by which budget totals can be found, nor any by which the budget can be viewed or treated as a whole. So there is no way that Lord Keynes's theories can be put to work.

Thus, in the fall of 1972, both houses of the Congress, acting separately, voted for a two hundred and fifty billion dollar spending ceiling. This was virtually unprecedented, but what followed was not: The ceiling measure itself was never enacted, because of disagreement in the conference, but it is instructive to note that less than two weeks after each house had separately voted for a ceiling of two hundred and fifty billion dollars, the Congress adjourned, having authorized, and, as they believed, directed, spending totaling some two hundred and sixty-three billion dollars. This is not pointed out in criticism of individual congressmen as much as it is in criticism of the lack of any coherent Congressional budget procedure. Given the fact that there is no opportunity for the Congress to consider the budget as a whole, it is always a matter of pure chance as to what is the total spending that will emerge at the end of a long session.

Another serious problem is that, for at least a generation, Congress has not completed its enactment of the budget within anything approaching even six months after the beginning of the new fiscal year for which the appropriations are made.

The Congressional problems, however, are really the second stage of the budget process, and they are mentioned first simply to show how far the best-laid plans of those who make up the budget can go agley.[5]

And they are well-laid plans. The beginning of the budget process usually starts in the spring of the year, before it is to be submitted to the Congress in January, and the budget that is submitted to the Congress in January is supposed to (but, as we have seen, rarely does) take effect the following July. So the budget process begins[6] approximately fifteen months before the time it should take effect, but, given the Congressional problems, about nineteen to twenty months before it is enacted. With the cloudiness that affects most crystal balls, particularly government-issued crystal balls, that in and of itself is almost enough reason why the fine-tuned Keynesian model envisioned by the theorists has little applicability to the facts.

In any event, in the early spring the Office of Management and Budget begins collecting the predictions of individual tax collections, business conditions, possible recessions; or unemployment rates, inflation rates, real growth, and all the other factors that are to be put into the models and produce the predictions of total revenue, and of the expansionary or recessionary nature of events that are supposed to begin unfolding twenty-two months away and will continue to unfold for twelve more months beyond that.

The raw data gathered by this process is then augmented by similar data or measured guesses being produced by the various other elements of the public, quasi-public, and private sectors, all of whom need a look, however cloudy, into the quite distant future. These include banks, investment houses, large corporations, Federal Reserve districts, and economic organizations which compete with one another over the accuracy of their prophesies. This is a business where a small error of a few percentage points can, of course, mean billions of dollars. Those small errors are made repeatedly, but because of the importance of having *some* kind of data with which to begin the process, past failures are either forgotten or tactfully ignored.

At the conclusion of various meetings culminating in gatherings of the Office of Management and Budget top staff with the director and deputy director, a figure, or rather a rough range of figures, is produced which the director can report to the President as being the probable revenues that will be collected, given no change in the tax system, and the further assumption that some or all of the predictions are reasonably accurate.

During the spring, other gatherings of the Council of Economic Advisors, the Quadriad, the Troika, and various individuals whom the President wishes to consult will produce opinions and recommendations, and out of all this will emerge not only the predicted revenues for the fiscal year that is to begin, now a little over a year away, but also predictions of the path of the economy during the twelve months of that next fiscal year.

At the same time, other groups in the Office of Management and Budget will have been gathering, or rather more accurately in the last few years, bringing up to date the expenditure data already compiled as part of what is called the five-year rolling budget estimates. This is a system inaugurated in 1970, under which the Office of Management and Budget attempts to predict in the roundest of numbers the expenditures that would be required to fund an assumed static state of existing programs, five years out into the future.

With some 75 per cent or more of the federal budget classified as uncontrollable—that is, about three quarters of the federal budget is virtually already made up with obligations under existing programs which will have to be honored, unless any of those programs are changed, it becomes increasingly important to know approximately what those obligations are. With three quarters of the whole federal budget in this category, theoretically at least about a quarter of the rest has some degree of flexibility, meaning some discretionary judgment may be exercised with respect to that fraction.

Therefore, at the end of the spring, the OMB seers

present not only estimates of how much revenue we will get, far out in the future, and what will then be the state of the economy, unemployment, gross national product, inflation, etc., but also approximately how much will be required to continue funding all of the existing programs, assuming no basic change up or down in the level of those programs (which is a pretty wild assumption), and given also the assumption that Congress will not terminate any of the programs (which is the only virtual certainty in the whole process).

But it should not be supposed that this remaining 25 per cent *has* any real degree of flexibility in it. Within that 25 per cent, we must spend most of whatever it is that we will spend on national defense, on public works, on space exploration, on bio-medical and other research and development, for any salary increases, and for unexpected contingencies which, remember, are a year and a half to two years away. Finally, within this total we must also include whatever new programs any President would like to propose even though they may not be yet drafted, nor have any cost estimates or even any chance of enactment.

Another reason why this theoretical 25 per cent of the budget is really not controllable is that *any decrease* from any program is always viewed as a retreat, a loss of commitment, or such serious backsliding as not to be tolerated.

Finally, at the end of this spring process a set of totals can be roughly seen that will generally indicate that more money will have to be spent than the anticipated revenues, without any new presidential initiatives or expansion room in the existing programs. This disparity is usually quite large even without allowing for "Congressional threats," meaning the possibility that Congress might enact expenditures even higher than those required by the then existing statutes for some of the uncontrollable programs.

Meanwhile, verifying hearings, called the "spring previews," are being conducted within the Office of Management and Budget. In these, the principal staff officers of the office, who, with their associates, have worked with the program managers in the various departments, present to OMB officials the probable path of most of the major spending programs of the government grouped by departments or, increasingly, by subject matter. In these meetings there are discussed possible changes of directions of the various programs and departments which could produce some significant improvement in the ability to reach the objectives of the programs themselves, or to free money for some new and more promising programs by stopping some existing ones. There is always an air of unreality about these discussions, unless the proposal can be accomplished without legislation, because everyone in the room realizes that the Congress is almost certain to reject any proposal that involves stopping anything the government is doing even if new proposals are made to take its place. These spring previews, however, do perform a useful and very necessary function. They not only give budget-making officials an opportunity to see where our present course will take us, but also an opportunity to review the effectiveness and usefulness of many of our programs. They also give departmental program managers an opportunity to indicate their plans for the future. Sometimes these objectives aren't reached, but at least a degree of forward thinking and planning is forced by the process that might well not occur otherwise.

From all of this comes a very rough schematic sketch of the next year's budget, including estimated revenues, estimated required expenditures by department, and the amazingly small percentage that is available for the President to propose be used for new programs or different directions. All of this material is then presented to the President by the director and deputy director of OMB, usually in the early summer. His decisions are given either in writing as part of decision memoranda, or growing out of occasional discussions in Cabinet meetings, or, more likely, discussions with the director of OMB, the Secretary of the Treasury, or, depending on the style of that particular President, other advisers. In the case of President Nixon, Domestic Council staff personnel were also consulted.

These decisions are then communicated in the form of ceiling letters to the various departments. These are letters advising the departments that in no event should their budget planning exceed the amount in the ceiling letter and that it may well be other events will require they receive even lower amounts, and that they should produce a budget proposal within that ceiling for the OMB and the President to consider by September. These proposals are then made up by the departments and considered by the OMB staff, considered again when the staff presents them to the OMB senior officials and the director, and considered a third time when they are presented to the President or his principal domestic advisers. The latter process happens with only a few of the major items, such as the amount of the defense budget, the spending that will be incurred for space or environment, or other principal issues of the day.

The process then proceeds mechanically with an increasing degree of automation and use of computers. The vast majority of the final budget decisions are those recommended by the OMB to the President in the early fall. A few major ones are reserved for further discussion, and occasional pleas by Department heads may reach to some of the assistant directors of OMB or in comparatively rare cases to the director or in even rarer cases to the President. Obviously, again, much will depend on the style of the President. President Nixon disliked budget appeals. President

Truman was reported to relish deciding them.

Thus, in many cases what began as a departmental request fitted within an OMB ceiling becomes in fact the actual budget item with perhaps little alteration. But an issue such as the total size of the defense budget, which is viewed both at home and abroad as an important signal as to the readiness and (even more important) the resolution of the country, may be delayed for final decision until as late as Christmas Eve.

Simultaneously with all this, the printing of some of the text and special analyses, most of which are made up more for the benefit of practicing economists than for the people, has been going on since early November. Incredibly, the whole massive production job in four volumes is nearly always ready for the Congress on schedule, fifteen days after the new session, in January, convenes.

Where in all of this process, you may now justly ask, is there either creativity or collaboration?

There is collaboration throughout the entire process: the interaction of the people who gather a number of strands of data and weave them into a coherent and complete prediction, even if not an accurate one; the process of consultation, interaction, and discussion within the Office of Management and Budget that has to occur each year; the same process that takes place between the OMB and the President's economic and domestic advisers, and between his advisers and the President; the discussions and debates between the departmental officials and the OMB; the hearing process both in the spring and in the fall, and the working together between budget officials, the economic and domestic advisers, and the President; and the ultimate attempts to win the decision of the President—all of this is the collaborative process in full swing. It is an example of collaboration restraining unbridled creativity.

The Budget Process

The fact that the ultimate results from it may not be the results desired by all, or indeed sometimes by any, is an inevitable by-product of such a process, and indeed an inevitable by-product of the budget system. For, as we have seen, our disposable resources are limited, despite the immense totals, and the discretionary authority of the President is strictly limited. Therefore, inevitably a *selection* must be made from among the many things that people would like to do. Whenever a selection process is involved, more are dissatisfied than satisfied.

Because of this, it is essential that the process involve as much collaboration as possible even though, ultimately, decisions do have to be made—and it cannot be all that "ultimately" after all. For there is one abiding characteristic of the budget process, and that is that it *must* be completed on time. The budget cannot be deferred until next year, or given to a task force to study, or postponed by any other of the favorite methods. The budget must reach the Congress on time, no matter how long the Congress may delay its enactment. It is this deadline which drives so many decisions and forces so many answers. It is inevitably a part of the governmental process that people would like to avoid. For most politicians and others prefer not to make hard decisions or give answers that disappoint and anger. In many parts of the governmental process this can be and is avoided, particularly in the legislative branch. But the budget must be finished by the fifteenth day after the Congress reassembles in January, and skilled as the Government Printing Office is (and it is extraordinarily skilled in performing its mechanical tasks), a certain number of days must be allowed for this skilled process to operate. So it is inevitable that thousands of vitally important decisions must be made in the fall, and in no event can any but a very few be postponed beyond Christmas.

Many will feel that there is little if any opportunity for any creativity to be exercised within such a large, complex, and in many ways so mechanical a process as that which results in the production of four large volumes less than eight months after the process begins.

To concede that, however, would be to condemn ourselves to sterile sameness year after year, and would also be to overlook many notably creative, innovative, and highly important changes that have occurred, and ideas that have been spawned within this increasingly complex process.

A few examples are clearly in order. The five-year rolling budget mentioned a moment ago is one of these ideas. As a result, each year does not take one completely by surprise, and a somewhat longer-range view is possible than the narrow, artificial twelve months of the fiscal year into which the actual budget must be compressed. When you spread out the fiscal requirements of each existing major program over five years, it is easy to be astounded at the growth that is mandated by each program itself, and also to be horrified at the percentage of the total available future activity of the government that is mortgaged by a single enactment in the past. Thus, each year we make loan commitments including a subsidy of, say, one billion dollars, we are committing ourselves to outlays of one billion dollars a year for thirty years or more out into the future. None of that was fully appreciated until each year we laid out expenditure estimates, five years into the future, for the increasing costs of subsidized housing, or environmental protection, or of maintaining an all-volunteer army with its built-in wage increases, or of funding bio-medical research on a same scale of increase as had obtained in the past. So, to those who are willing to think the unthinkable—namely, changing some of the existing programs, the five-year rolling budget, in which each year new figures are posted for year five, and each year adjustments are made for year one, was a creative new idea of great importance.

Many of the most creative ideas in government have come from bio-medical research or the space and the atomic energy programs, in which new frontiers of knowledge, of exploration, of potential new resources, and of solutions both to energy and defense problems, beckon men of vision and of daring. One in particular involved considerable creativity on the part of the budget officers. This was the ultimate decision to fund the space-shuttle program. The space shuttle is a reusable set of space vehicles that would be boosted into space, perform their tasks, and come back to landings that would enable them to be used again. In 1971, the NASA proposal was for a $9.2 billion, six-year shuttle proposal.

There were protracted discussions between NASA officers, potential designers, and the OMB's budget and evaluation people. Many of the latter did not want to fund it at all; some said that what NASA, and the country, needed could be done less expensively and more effectively. The first decision, therefore, was that we would build the shuttle. Even though its precise use could not be spelled out, the promise of a system of reusable space vehicles that could go up and perform numerous tasks and return with soft landings permitting unlimited future uses, proved irresistible. In an age of space, the nation clearly has a future in space and that future can best be realized by investing now in a system which could launch vehicles at about $10 million per launch rather than the hundreds of millions per launch now required.

Once the basic decision was made by the director of OMB to recommend a shuttle to the President, then the original design was re-examined by NASA with OMB prodding, and a new design was proposed. The shuttle system finally recommended by OMB and NASA, and approved by the President, calls for spending $5.1 billion over the next six years, which was nearly a 50 per cent reduction in the original NASA proposal. This cost cutting was accomplished by reducing the size of the payload the shuttle could carry and accepting the fact that smaller payloads with a reusable system were quite acceptable, in view of the number of trips that could be made and the higher degree of miniaturization normally to be expected in the years ahead, when the shuttle will be ready. The original $10 billion design was so costly it would virtually prohibit construction—but a collaborative exercise, marked by creativity on all sides, produced a new design and saved this vital key to a part of our undoubted future.

In the fiscal 1974 budget, another example of creativity was seen: The problem was to try somehow to convince the Congress and executive-branch departments that the total expenditures of the United States Government were virtually out of control, and that the 75 per cent "uncontrollable" portion of the budget was more than just a phrase. The difficulty was not only that the Congress had no procedure for looking at the budget as a whole, but that there was far too little appreciation of the long-range requirements of individual programs. This was best seen by demonstrating, as could easily be done, that if the budget for fiscal '73 were held to approximately $250 billion of total expenditures, the next year's budget, with some but not undue restraint, could be held to the $268 billion range, and that the budget for fiscal '75, again with some restraint, might be expected to reach $286 billion or a little less.

However, if the $250 billion ceiling were not achieved in fiscal '73, and actual spending reached the $263 billion Congress had appropriated when it adjourned in December 1972, then the fiscal '74 budget, instead of being a total of

$268 billion, would automatically climb to somewhere in the neighborhood of $285 billion even without any new additions. Given this same pattern, the fiscal '75 budget would be about $315 billion instead of the $286 billion which could be achieved if we held to the $250 billion in 1973.

These high figures for 1974 and 1975 were far ahead of the predicted revenues and convinced most economists that we could not spend at anywhere near those projected levels without incurring the most damaging inflation. The problem was that no one in the Congress really believed this, or, more accurately, wanted to be bothered with the demonstration that inflation was almost certain to follow unrestrained spending.[7]

The solution, therefore, was not to rely on such comparatively academic and remote concepts as the five-year rolling budget; rather, the solution was to print three years' budgets in very considerable detail in the volumes that normally would have contained just the 1974 budget. We brought up to date the 1973 budget, which had been submitted the preceding January, incorporating all of the Congressional changes and the most likely Congressional threats still to appear; we submitted the regular 1974 budget in the amount of $268 billion, which was based entirely on the assumption that we would be able to stay within $250 billion for fiscal '73; and, for the first time, we printed a quite detailed fiscal '75 budget showing that it could be held in the $287 billion range rather than the $315 billion which would be compelled if we let things loose in 1973 and spent all Congress appropriated.

The dynamism of federal spending programs, however, is such that the only way that we could achieve the $268 billion recommendation for fiscal '74 was to meet head-on the Congressional threats for more fiscal '73 spending, and to set up firm executive barriers to the normally unstoppable onward rush of expenditures for fiscal '74. There were only two ways in which this could have been done: One was to make recommendations to the Congress not to spend, which we did with the usual feeling of hopelessness. The other was to exercise the power that every President since Thomas Jefferson has exercised—and that is *not* to spend all that was appropriated for various given programs. We used both methods, and for fiscal '73 we succeeded. The actual total spending for that year was brought in under the $250 billion mark—approximately $246 billion in fact.

This would have enabled the fiscal '74 figures to come in at approximately $268 billion—the amount the President recommended—if the Congress had followed his recommendations. The important thing to note is that this was the *only* way in which we could possibly have achieved a total of $268 billion for '74, because if we had allowed fiscal '73 to run its normal course (normal, that is, by prior years' experience), expenditures that year would have been so high that, with three quarters of the total budget uncontrollable, the next year—fiscal '74—would inevitably have been far above the projected figure and far above the revenues, and clearly inflationary.

The creativity or novelty of the fiscal '74 budget lay in the fact that it projected for the first time the full effects of this ongoing dynamism of federal spending; showed precisely where it would lead if unrestrained in 1975; and attempted, thereby, to convince the Congress that the normal path should not be trod. To the extent that it failed to persuade the Congress, this new idea can be said to have been a failure. But, to the extent that it helped force the Congress for the first time in many years to do some serious thinking about its own budget procedures, it can be said to have been a partial success.

How effective will be Congress' self-examination of ways to improve its consideration remains to be seen. The new budget-reform act means we will get a new committee, a new staff, a new fiscal year, and a new timetable. But there is nothing in the new bill which guarantees either enactment of a budget on time, or a willingness to stay under any previously set ceiling. In fact, the bill specifically allows Congress to raise its ceiling any time it wishes.

The real difficulty, of course, is that the Congress, and indeed many of those in the executive branch, are quite unwilling to face the inescapable, simple logic that tells us that the only way to reduce the total, or even to hold the total within reasonable limits, is to act upon the thirteen or fourteen hundred separate pieces that make up the total. And it is this examination, program by program, plus the willingness to reduce or eliminate or substitute, that the Congress seems unwilling to do.

Large sums are usually appropriated by the Congress to evaluate various programs. The executive branch and the Office of Management and Budget devote substantial amounts of time and resources to the same purpose. Everyone agrees that each program should be carefully evaluated. Indeed it is hard to argue the point that we should know how well individual programs are working and whether they are achieving their goals—assuming that the goals can be stated. But this is a totally meaningless exercise unless somebody is willing to act on the information produced. The executive branch has acted on it from time to time, but usually the Congress refuses to abide by any evaluation action which involves stopping or reducing a program.

For example, the Administration learned through evaluation that federally subsidized housing programs for the most part benefit middle- or upper-income people and builders and developers. We also learned that, each year, an increasing amount of our future was mortgaged, in that each year's expenditures for federally subsidized housing meant years of payments out into the future for each project. The Administration proposed terminating these subsidies, and the resulting storm in Congress was comparable to a tornado. We reported to the Congress that there was a substantial surplus of hospital beds in the country and that it was foolish—not only foolish but quite seriously wrong—to continue subsidizing more and more hospital construction under the Hill-Burton Act. The more Hill-Burton funds were used, the greater the number of unused hospital beds that would be produced. It was readily apparent that the cost of operating hospitals with empty beds is vastly greater than with full occupancy, and that this unnecessary increased cost would cause an inflationary rise in hospital costs throughout the country. We proposed not spending the funds, but Congress and the numerous private interests involved went to court, and the trial-court decisions so far are that the President could not impound for such a purpose. While appellate courts have not yet reached such a result, we are meanwhile back subsidizing more empty hospital beds.[8]

Since the whole budget process involves the *selection* of programs on which to spend the available resources, and since the selection process necessarily involves funding some programs, and reducing or terminating others, evaluation is meaningless if the ultimate answer is to be that nothing will ever be terminated or reduced. But that is the answer that Congress, and even the executive branch on occasion, has repeatedly returned.

In the State of California we were able to show somewhat better results.

I recall, when I was director of finance of California,

budget hearings which showed that juvenile delinquency increased regularly, matched by juvenile arrests, convictions, and sentencing to detention homes, or, as they were gently called, Youth Authority Facilities. But the more confinement, the more juvenile crimes seemed to take place and the higher the rate of recidivism.

It was noticed that only a small percentage of the total Youth Authority juvenile budget went into research to develop basic causes and attempted cures. When this was noticed and compared with the demoralizing and increasing costs of arrests and detentions, and the lack of positive results, a switch in funding was made with an increase in research, which, in turn, caused a shift of emphasis within that department.

The budgeting procedures under which these facts could be noticed and priorities shifted was called program budgeting. That, too, represented a creative idea, though shortly it became so overlaid with jargon and professionalese that only a few could pretend to try to understand or apply it—which, of course, was the reason why jargon and professionalese are used in the first place. Essentially, program budgeting consists—or should consist—of allocating funds for various programs rather than by departments. This involves a statement of the objectives of the programs by the managers. For that reason, it was bitterly opposed by many program managers who found themselves quite unable to write a simple, clearly understandable statement of what it was they were trying to accomplish and why—in short, what it was they were doing. But, once this difficulty was surmounted, the task of allocating resources among programs was not only easier, but far more effective. We could then measure progress toward the stated objective and fund accordingly, and we could also help determine whether progress toward that objective was indeed useful.

Program budgeting in this simplified form was installed in California in 1968 over reasonably strenuous opposition, and it has been working well since. One short example of how it worked could be seen in that instead of simply considering a budget for the Department of Motor Vehicles, the requests for the individual programs that department was administering were examined to determine how well those programs had worked. One of them involved driver-training schools for persons arrested for what are called moving violations. The program managers set out with considerable pride the increasing amounts that had been spent on driver-training schools for persons with three or more moving violations, and they requested even more for the following year. We asked what were the records of those who had been through the driver-training school. That question effectively brought the requested increase to a halt. It was learned, to everyone's astonishment, that graduates of the driver-training school went on to have even more violations *after* they had been to school. I did not stay long enough to find out if their records began to improve when they were no longer sent to driver-training schools—but I'm entitled to my suspicions.

All attempts to improve the budget process, or indeed to understand it, go back to a very fundamental and very unfortunate factor in American public and private life: virtually everything is judged by the amount of money that goes into it. Fundamentally, that is the reason why so many Congressmen—far more than a majority in almost every case—are unwilling to reduce or eliminate programs, or indeed even to moderate the rate of growth of many programs. For they believe, and they are probably right, that their commitment to any cause will be judged by their constituents (or at least the vocal ones) solely in terms of how much they are willing to spend on that cause.

Thus, if bio-medical research, which is clearly an exceptionally important and necessary governmental activity, is not increased at least 25 per cent each year, a number of people who are otherwise quite rational in their thinking will immediately argue that those opposed to such an increase are against medicine, are in favor of disease, and are totally unworthy of serving a great and healthy public. The fact that such a rate of increase, in one item, coupled with similar rates of increase demanded by advocates of all of the other programs would totally bankrupt the country, or would require such a high rate of taxation that the free-enterprise system which produces the incomes necessary to pay such a rate of taxation would be virtually destroyed, or would cause a rate of inflation totally unsupportable to everyone, is considered not relevant to the discussion.

Given this attitude and the fact that the Congress, and indeed far too many in the executive branch, are influenced primarily not by "the people" but by interest groups, most of which are based in Washington, and all of which want a far higher rate of increased spending for all of their programs, and view with the unmitigated horror that one views personal unemployment any suggestion that programs in or for which they work might be reduced or terminated, it seems inevitable that this headlong increase in governmental spending, activity, and power will continue until the government uses such a large percentage of the total gross national product that a free-enterprise system will be virtually extinct.

The only thing that would appear possible to break the cycle, and this will require perhaps the greatest creativity of all, will be to bring home to enough of the people that this is the inevitable end of allowing only those forces for increased governmental activity, spending, and power to hold the field.

For, essentially, this is the problem. Virtually the only people that the Congress hears from and the only people who attempt to influence the executive branch are those who are united in wanting programs continued. Lobbyists, staffs of Congressional committees which oversee the programs, and those permanent employees of the government who administer them—these three groups have been appropriately called "the iron triangle." In combination they are far more powerful than occasional lonely voices raised against either individual programs, or the trend caused by the predominance of the iron triangle.

Those crying out thus in the wilderness do not want all government to come to a halt. They seek a careful examination of the objectives of each program, and an answer to the question as to whether the administration of the program at last year's level of funding helped us achieve an objective that is important, necessary, and beneficial for society to achieve, and one that could have been achieved only by the government. Most important of all, they seek a willingness to stop doing what we are doing if it appears that we are not achieving a desirable or a necessary objective with that program.

The point of all this is not just to save money, or to reduce the size and power of government (although these are not undesirable objectives). It is to free available government resources for better use, and thereby halt the trend of simply funding every new idea as an *addition* to the total funds being expended for all of the old programs, some of which, at least, it can be demonstrated do not work or are no longer needed.

In short, there must be enough people willing to look each existing program coolly in the eye and ask if it is working or if the federal government, rather than the private sector or some other level of government, has to do it, and if not, why it should not be stopped so that the money can be used for a more useful and effective purpose.

This will take both creativity and collaboration. The collaboration within the budget process is clearly there now. That entire process, as we have seen, is a collaborative effort, a pyramidal structure, culminating in various decisions made at various altitudes up the pyramid, with the most important saved for the President.

This process results, then, in the budget, which is, as we have seen, simply a set of proposals or recommendations. The pyramid is then taken apart in the Congressional process by a vast number of separate groups who do not look at the pyramid but only at each of the stones. The resulting product is not a pyramid, nor indeed any recognizable shape, but is a combination of the total demands of the most vocal and effective small interest groups.

The creativity that is exemplified and needed in the course of this process is every bit as vital to the continuation and improvement of society and the quality of life as is creativity in scientific and artistic processes.

The difference is that, thus far, creativity has been virtually swamped by the basic imperative of judging everything by total amounts of money rather than results, necessities, and progress. The ultimate creativity in the budget process will come only when we have turned the public's mind to judging by results and actual necessities, rather than by totals.

The Frontiers of Knowledge

FOOTNOTES

[1] "The Affective Domain and Beyond," by Dr. Robert K. Kantor, Stanford Research Institute, 1968, Research Note EPRC-6747-8.

[2] Mayo Pines, *The Brain Changers—Scientists and the New Mind Control* (New York: Harcourt Brace Jovanovich, 1974). Einstein said he arrived at his ideas without thinking "in words."

[3] "The Brain's Generation Gap," by Paul MacLean, Chief of the Laboratory of Brain Evaluation Behavior, NIMH. A Lecture, AAAS, December 28, 1971.

[4] Because so many *do* think so, it tends to become a self-fulfilling view, because the economy *is* greatly affected by how things are perceived—whether they are correctly perceived or not.

[5] The new Budget Reform Act of 1974, effective in 1976, may change some of this for the better—see page 259.

[6] Actually, as we will see with the five-year rolling budget concept, the planning begins far earlier.

[7] "Almost certain" because the public perceives unrestrained government spending as causing inflation; and what the public perceives frequently comes to pass, whether the perception is correct or not.

[8] The Administration was ordered by the U. S. District Court for the District of Columbia to allocate the impounded Hill-Burton Act funds, a decision it decided not to appeal.

THIRD SERIES

THE MODERN
EXPLORERS

1974–75

SIR EDMUND HILLARY

SIR FRED HOYLE

WILLARD F. LIBBY

ISAAC ASIMOV

SOUTH POLE—
CONTINENT OF ADVENTURE

SIR EDMUND HILLARY *Sir Edmund Hillary's explorations have taken him to the summit of Mount Everest, the world's highest mountain, and to the earth's southern pole. On May 29, 1953, the New Zealander and his Sherpa guide Tenzing Norgay reached the peak of Everest, a height of more than 29,000 feet—the first climbers to make a successful challenge.*

In 1958, Hillary led the third party to reach the South Pole overland. Preceded by Amundsen and Scott almost fifty years before, Hillary and four other New Zealanders made a twelve-hundred-mile tractor journey from Scott Station, on the Pacific coast of Antarctica. Hillary's party had joined a British effort during the 1958 International Geophysical Year to make the first surface crossing of the Antarctic Continent. After establishing necessary fuel depots along the Ross Dependency for Sir Vivian Fuchs's crossing party, Hillary and his men drove the additional five hundred miles to the Pole, reaching it two months before Fuchs.

I

I REGARD it as a considerable honor to have been invited to address such a redoubtable organization as the Smithsonian Institution, but I also feel a certain sense of guilt because in a way I'm here under false pretenses.

Some years ago, I received an invitation from the Vice-Chancellor of the University of Natal to deliver an address at (and I quote) "an important conference of distinguished scientists and educators." I wrote back to the Vice-Chancellor and said that quite clearly he had mistaken me for someone else, as I was neither a distinguished scientist nor a distinguished educator. I almost sent a similar letter to the Director of the National Museum of History and Technology.

These lectures are being delivered largely by eminent scientists, in one of the most famous scientific institutions in the world. Although I like to regard myself as a bit of an explorer—and even an innovator—I am, alas, no scientist, but merely an enthusiastic layman.

Over the years, substantial scientific work has been undertaken in the Antarctic, often under conditions of considerable discomfort. There are probably few places where science is so dependent on people who have an adventurous spirit and an instinct for survival.

But I don't intend to talk about science. I plan to concentrate on the things I know best—exploration, re-sourcefulness, logistics, and the joy of meeting and over-coming a tough challenge.

I have often resented the way that science has sometimes been introduced to justify an interesting adventure—particularly if quite a lot of money is required. Adventure is worthwhile for its own sake. How many of us have been stimulated by some glorious effort that had no conceivable economic or scientific reward?

But in the Antarctic (perhaps more than in most places) there has over the years been a happy and fruitful co-operation between the explorer and the scientist.

II

Two hundred years ago, the world must have been a great place for an explorer: the West had still to be won; Australia was inhabited by aborigines and kangaroos; much of the globe wasn't mapped. And no one knew that Mount Everest existed. But for centuries—ever since Aristotle propounded a spherical earth—cartographers and geographers had been surmising that a great southern continent existed, "Terra Australis," whose bulk would balance the land masses of Europe and Asia in the Northern Hemisphere. Most of them thought of this area as being rich in gold, elephants, and valuable spices.

Abel Janszoon Tasman was seeking the fabled Terra Australis when he discovered Tasmania and New Zealand in 1642, but it was left to the redoubtable British navigator Captain James Cook, over a century later, to define the extent of the southern continent. In his great expedition of 1772–75, he probed the Antarctic ice edge. On the seventeenth of January, 1773, he crossed the Antarctic Circle for the first time in recorded history and later crossed it several times more, reaching 71° south. It was an astonishing effort in a small sailing vessel half a world away from home. In his voyage around the Antarctic Continent, Cook never penetrated through the pack ice to solid land, but he had at least proved that if a continent did indeed exist it must lie in a region of perpetual snow and ice—and that it was unlikely there would be too many elephants and spices.

The first Antarctic explorer may well have been a Polynesian. Maori legend tells of the chieftain Ui-te-Rangiora, who in A.D. 650—about thirteen hundred years ago—journeyed far south until he saw what he called "moving white cliffs in the sea of *Pia*." *Pia* means arrowroot and affords a good description of brash ice. It is not beyond the bounds of possibility that the Polynesian warriors who sailed the vast reaches of the Pacific in their outrigger canoes might have strayed south to the Antarctic Circle.

Who in fact did make the first recorded sighting of the Antarctic land mass? There has been much disagreement among historians, who often appear to have been motivated by national loyalties rather than by an unbiased search for the truth. Was it Britisher Bransfield in the *Williams* or American Palmer in the *Mer?* Or was it indeed, as there is strong evidence to indicate, the great Russian navigator Bellingshausen? All these men had sightings during 1819 of what could indeed have been the main continent, but bad weather and the usual poor visibility made confirmation difficult.

It seems more than likely that many sightings had been made before this by the hardy sea captains from New England who for some years previously had been pushing relentlessly southward seeking their bloody harvest of fur seals. These tough sailors were much to be admired, even if their actions almost brought to extinction the vast herds of helpless seals. (How can *we* complain? Even today, we accept the same ruthless destruction of the mighty whale.)

Who made the first landing on the southern continent? The renowned explorers Dumont d'Urville, Wilkes, and Ross have all claimed this honor, but once again it seems reasonable to suggest that it was some enthusiastic seal hunter—although few of them left useful records. They were businessmen in a very competitive field, and any new discoveries or landings would not be reported lest other sealers take advantage to their own profit. Mr. Ian Cameron, in his recently published book *Antarctica—The Last Continent*, quotes from the log of Captain Davis of Connecticut on the vessel *Cecilia:*

Wednesday 7th February 1821
Commences with open cloudy weather and light winds a standing for a large body of land in that direction S.E. at 10 A.M. close in with it, out boat and sent her on shore to look for seal. At 11 A.M., the boat returned but found no seal. At noon our latitude was 64°01 South. Stood up a Large Bay, the land high and covered entirely with snow . . . I think this Southern Land to be a continent.

This could be the earliest recorded account of actually setting foot on the continent, and so in the interim at least the honor should, possibly, go to the little-known Captain Davis of New Haven, Connecticut.

In the period around 1840, three great nationally sponsored expeditions investigated the fringes of the continent. J. S. C. Dumont d'Urville led a French expedition with two ships. The United States Exploring Expedition (a flagship and five small vessels) was under the leadership of Charles Wilkes—a somewhat quarrelsome yet daring explorer who became generally admired in Europe yet was reviled and subjected to a court-martial in his homeland.

And then there was Britisher James Clark Ross—experienced, competent, and well supported. His two vessels, *Erebus* and *Terror*, were carefully chosen for the task and specially strengthened to withstand ice pressure. With courage and determination, he pushed southward through the pack but barely survived a most appalling storm, which put his ships in grave danger from jostling icebergs. Early in January 1841, he burst through the pack into the open sea that now bears his name—the Ross Sea. Sailing onward, he saw the high mountain ranges of Victoria Land and then turned eastward and sailed along the edge of the mighty ice shelf that has also been named after him. The active volcano Mount Erebus was also one of his discoveries.

As a native of New Zealand, a member country of the British Empire (as it was then called), I was brought up on a diet of the heroic efforts of the Antarctic explorers at the beginning of the twentieth century. The epic adventures of Nordenskjöld's party during their two winters on the Antarctic Peninsula largely passed me by—they achieved little in latitude in their push for the pole; but they gained vast experience in Antarctic living and traveling, and they produced substantial scientific results.

Almost at the same time, Captain Robert Falcon Scott took a British expedition (1901–4) into the Ross Sea and wintered over in McMurdo Sound, a place that was to become very familiar to me in later years. During the course of this expedition, Scott and his men traveled up the Ferrer Glacier and reached the polar plateau west of Ross Island. They laid several depots to the south, and a three-man party composed of Scott himself plus Dr. Edward Wilson and Lieutenant Ernest Shackleton sledged southward over the Ross Ice Shelf. They surveyed and named the new peaks that almost daily came into view until their dogs almost failed and progress became pitifully slow. Always to the south stretched endless miles of flat, snowy expanse. Finally Dr. Wilson discovered that Shackleton was suffering from scurvy and return was vital. Also, their supplies of food and fuel were getting very low. In fifty-nine days they had covered only 380 miles—averaging little over six miles a day—but they had established a record for farthest south. On the return journey, all the remaining dogs died and it became a race to reach safety before their supplies failed. It took them thirty-four days to reach the ship and safety, and Shackleton was then invalided home.

Shackleton was most disappointed at having to withdraw from the expedition, and he resolved to return with an expedition of his own. He was an amazing character. Born in Ireland of Irish/Yorkshire parentage, and with a colorful and flamboyant disposition, he combined daring with caution, and a love of poetry with the ability to knock a man down in a roughhouse. He was certainly my hero as far as Antarctic exploration was concerned. He didn't bother with an organizing committee but set to work to raise the funds himself and took responsibility himself for the thousands of decisions necessary to launch an Antarctic expedition.

There were many ways in which Shackleton deviated with marked success from earlier expeditions: he chose a land party distinct from the ship's company; he was the first to take a motor vehicle to the Antarctic; he was the first of all polar explorers to use acetylene gas to lighten the winter darkness of his quarters. Shackleton's rations were luxurious to an extent thitherto unknown in such expeditions (I ate some of them fifty years later at his winter headquarters at Cape Royds). This diverse and generous food ration was one of the primary causes of the great measure of success of his expedition as a whole. Shackleton personally chose and tested each item and supervised its packaging. For the South Pole journey, he chose Manchurian ponies to haul sledges that could be man-hauled if the ponies failed. Camping methods, sledge handling, and marking of the trail were upgraded, and this enabled a considerable improvement in their speed.

Shackleton and his men left New Zealand on New Year's Day 1908 in the *Nimrod*. In McMurdo Sound, the fast ice was much farther north than in 1902, and a new camp was therefore set up at Cape Royds, about twenty miles from Hut Point. All was hustle and bustle until the hut was erected and the *Nimrod* sailed away.

Mount Erebus, 13,350 feet, the active volcano on Ross Island, was climbed by the party on 10 March 1908. This was an auspicious beginning to the expedition, but disappointment soon followed. Normally, the sea ice would have allowed members to sledge to Hut Point with ease and enable them to start laying depots southward across the Ross Ice Shelf. However, that year the ice broke up and floated out to sea shortly after their vessel had left for New Zealand, and they were penned in at Cape Royds. This meant a late start the next summer and no depots in place.

Despite this grave handicap, Shackleton and his men did magnificently. One party explored the western mountains of Victoria Land; another reached the South Magnetic Pole for the first time; the third, under Shackleton himself, struck southward toward the geographic Pole. Their ponies hauled the sledges at first, but they were then shot for food and the men took over the sledge hauling themselves. With every mile past Scott's farthest point, they opened up new mountains, but gradually the ranges tended eastward and crossed their path. When they could veer east no longer, they turned south again, and to their relief found a highway opening up before them. But what a highway it proved to be: the hundred miles of turgid ice that makes up the giant Beardmore Glacier(!) I have flown up this glacier and marveled at how Shackleton managed to pick a feasible route through the contorted pressure ridges and maze of crevasses. However, up it they duly went and then out across the polar plateau at a height of approximately ten thousand feet. Only ninety-seven miles from the Pole, they dared go no farther, for food was too short. The Pole could have been reached, but they would not have won back to safety. Shackleton had to make the heart-rending decision to return, and the race back to McMurdo Sound was a race against death itself, so short was the supply of food. Upon their return to civilization the world acclaimed Shackleton as a hero, and his journey was regarded as the greatest sledging trip ever undertaken, as it well may have been.

In 1910 Captain Scott was back again with a dash for the Pole as the major objective. But formidable competition was already in the field—Roald Amundsen, the very competent Norwegian explorer. Amundsen had plenty of husky dogs and a small team of expert skiers and dog drivers. In a brilliantly planned and executed tour de force, they pioneered a new route up onto the polar plateau and reached the South Pole on 14 December 1911. It was a highly professional performance. Scott meanwhile was having much greater problems: his ponies died one after the other and they were forced to man-haul; the Beardmore Glacier proved a terrible problem; and they arrived at the Pole on 17 January 1912—to see the Norwegian flag already there ahead of them. It must have been a terrible blow to their hopes and their morale.

Their return journey was even tougher. Rations were short, fuel scarce, and scurvy rampant. Evans died coming down the Beardmore Glacier. Oates walked out of the tent to his death so that he would no longer slow the progress of his comrades. The terrible cold sapped their strength and progress became alarmingly slow. Eleven miles from the plentiful supply at One Ton Depot, they pitched their last camp. Day after day of blizzard kept them there while supplies of fuel were exhausted. As their condition deteriorated, they still kept up their diaries and Scott wrote his famous letters, which were found months later beside his body. They had lost the race and death was their reward—but they became famous figures in British history because of this gallant failure.

With the reaching of the Pole, it might have been expected that there would have been a lull for a time in Antarctic exploration, but Shackleton had conceived an even more daring idea—an attempt to cross from one side to the other of the continent, through the South Pole. Alas for his ambitions, his ship was crushed in the heavy Weddell Sea pack ice and sank. With brilliant leadership, he extricated his party from this desperate predicament, bringing them safely over hundreds of miles of rough pack ice to the ocean. Shackleton sailed to South Georgia for help across the wild southern oceans and so performed one of the greatest small-boat journeys of all time.

The modern age of exploration started in the Antarctic in 1928 with Admiral Richard E. Byrd. In America the name of Byrd became synonomous with the Antarctic, although he never, perhaps, achieved the same sort of glamour elsewhere. He was handsome and adventurous, had great family and political influence, and was a successful and determined fund raiser. He introduced mechanization to the Antarctic—aircraft, radio, increasingly powerful technology. I think at times even Byrd felt that technology was coming between himself and the southern environment. Maybe this was why he wintered alone on the ice in 1934 with only his dog for a companion—a gesture, perhaps, to the heroic age. Byrd of course achieved great fame and prestige. I talked to him in 1956—not long before his death—and although he was in name still over-all leader of the U. S. Deep Freeze operation, the power had largely been taken out of his hands. He spoke wistfully of the past but was kind and encouraging for my own imminent projects. I sensed his wish to be with us, though he was now beyond it physically. The wish was there, perhaps, but the power had gone, and Byrd sounded like a sad and lonely old man. I guess we all come to that in the end.

The year 1953 was an eventful one for me. I climbed Everest, got married, and started a world-wide lecture tour—enough for anyone all in the same year. I was returning to the Himalayas in 1954—and after that? There was lots to do. Then I met Dr. Vivian Fuchs. Fuchs had plans to complete what Shackleton had set out to do—to cross the Antarctic Continent. I can clearly remember his rather scruffy office, his square, powerful figure, his complete conviction about the journey and its scientific worth. He would need a support party to go south from New Zealand and put depots out toward the Pole. He suggested I might be interested. I couldn't understand why he should be bothering with me—and then I realized that Fuchs hoped for vigorous support from New Zealand and I could help with this.

I didn't feel I had much in common with Fuchs—his interests and approach were completely different from mine—but over the next few years, I became more and more involved, and increasingly enthusiastic about the prospects for a New Zealand expedition based on McMurdo Sound. The International Geophysical Year was about to occur, so our program became increasingly complicated with scientific projects.

In December 1955 I sailed south with Fuchs in the

900-ton *Theron*—south from Montevideo into the Weddell Sea. Our objective was to lay the groundwork for the main crossing of the continent—to choose a site for Fuchs's base and to reconnoiter by air the access toward the South Pole. It was incredibly exciting to see my first iceberg and then to sail into the pack ice. Over the next month, I saw more than I wanted of the pack: we became jammed in heavy ice and it took us many weeks to shake ourselves free. We made our way to the bottom of the Weddell Sea and unloaded our supplies on the Filchner Ice Shelf. We had problems with storms and vehicles and the lateness of the season. When heavy pack ice approached, we sailed hastily to the north, leaving a small wintering party behind to build a base for the following year. I felt our party had been enthusiastic, but amateurish. There had been much for us to learn—and we'd learned most of it the hard way. At least we were now better prepared for the main journey the following year.

On my return to New Zealand, I was determined that my expedition should be as well trained and equipped as finance and time would permit. Although our support tasks to the crossing party were still our major responsibility, we had increasingly grown into a New Zealand national expedition, financially supported by the New Zealand government and by the business and private sector. Our scientific role as part of the International Geophysical Year program had become more and more important, and we were also planning a wide program of exploration, mapping, and geology. All members of my expedition took their turns at running with the dogs on the New Zealand glaciers, at driving tractors in the snow, at camping and living in Antarctic tents.

In our support role to Fuchs, the over-all plan had us laying depots several hundred miles toward the Pole, using dog teams and small aircraft for this purpose. Our few farm tractors were to be used for unloading the ship and for work around base—nothing else. But I already had ideas of taking vehicles farther south, toward the Pole, perhaps? We couldn't afford the powerful Sno-cats to be used by Fuchs's crossing party. Admiral George Dufek was lending us a couple of Weasels, which were effective but mechanically unreliable. If we wanted to use vehicles, we'd just have to use our farm tractors. I made sure we had enough of them, and a couple of resourceful engineers to keep them going.

My expedition was, I believe, a well-balanced one—a happy compromise between the old and the new. I suppose every leader thinks this! Our base, our scientific equipment, our communications—all were modern and efficient. In my field parties were men who had proved themselves in the mountains and on the glaciers, good hard men who could handle the problems of snow and ice and cold and live comfortably in tents at extreme temperatures. Our dog teams were strong and well trained. Our two small aircraft were well suited to their polar tasks. Only our farm tractors were inadequate for the job, although they did have the virtue of reliability in all weather and temperatures. In any case, they were all we had, and I was determined to make the most of them.

Our small wooden ship, the *Endeavour*, rolled steadily southward through the "roaring forties" and the "screaming fifties." At first the weather was surprisingly calm and even balmy, but then we had a wild and furious storm that had most of us stretched out on our bunks, with waves breaking over the dog kennels lashed on deck. It was an exciting moment when we saw the first piece of brash ice bobbing around in the waves. We passed through a line of icebergs and soon the pack ice was stretching from horizon

281

to horizon. We had no difficulty in breaking through at first, but gradually the ice thickened, and periodically we were forced to a halt. When the pressure eased, we would start southward again, pushing a narrow channel for hundreds of miles.

We broke out into the open waters of the Ross Sea and there wasn't a piece of ice in sight. It was almost like being in the open ocean, until the great bulk of Mount Erebus loomed up on the horizon with a long plume of smoke streaming from its summit. As we entered McMurdo Sound, we struck heavy pack ice again, and from then on there were plenty of decisions to make and quite a few headaches too. The site we planned to use at Butter Point proved completely unsuitable for our purpose, and in the end I chose Pram Point on Ross Island, facing south over the great Ross Ice Shelf.

We pulled in to the ice edge at McMurdo Sound and started the energetic task of unloading our hundreds of tons of supplies. Our dogs were staked out on the ice, very lively and boisterous—and glad to escape from the wet, confining conditions on the deck of the ship. We had a welcoming committee of giant emperor penguins: incredibly distinguished-looking birds, but also incredibly stupid. Swimming offshore were the killer whales, waiting patiently for a penguin or a seal—or a human—to come within their reach. We always kept one eye cocked in their direction: we were well aware of the reputation these whales have for tipping over heavy pieces of pack ice in order to throw any passengers into the water.

Twenty-four hours a day, our five small tractors relayed backward and forward across the bay ice, dragging loads of hut sections, food, fuel, and scientific equipment. Before long, Pram Point had become a hive of industry, like any construction site. We established a tent camp, and the construction party lived there while we prepared foundations for buildings and drilled holes into the permafrost to anchor tie-down rods. Our construction group were very expert; they had completely assembled and then disassembled the whole base back in Wellington, so the buildings went together at great speed. We could prepare the foundations one day and then the complete shell of a building would be erected and tied down the following day. Before long, we had plenty of shelter whatever the weather might do to us. Below the base, the Ross Ice Shelf ground its way slowly into the thick bay ice, forming high-pressure ridges. Hundreds of Weddell seals emerged through the cracks and had their pups on the ice.

Our two small aircraft were assembled and became operational, flying reconnaissance and supply missions. We had established a small airfield just below the camp, but our air crews had to keep very alert. If the bay ice had shown any signs of floating out to sea, the aircraft would need taxiing to safer ground or even have to take off and then land in a safer place on the Ross Ice Shelf.

With the departure of our ship, life at Scott Base became a race to get everything done before the onset of winter. There were supplies to be carefully stored, buildings to be completed, and journeys of reconnaissance to be undertaken. By the time the winter came, our dog teams had established a route up the Skelton Glacier onto the polar plateau, and we had a fully stocked depot 180 miles away from our base. When the darkness came, we were ready for it.

The long Antarctic night passed quickly by. We had few of the traditional problems associated with darkness and isolation: The twenty-three of us wintering over were far too busy carrying out the basic scientific program and making our preparations for the summer season. We had plenty of variety in our food, and all of us took a hand with the chores and even shared some of the cooking. Our communications with the outside world were excellent, and I could usually speak to my wife on the telephone once a week. Perhaps the improvement in communications was the most notable change between our lives and the lives of such explorers as Scott and Shackleton fifty years before. Our living room was bright and cheerful, and we took care to keep it that way. It was our home away from home, and we spent many happy and boisterous hours there. We had an adequate library, a record player plus a rather limited collection of records, and sufficient alcohol to have a good party once a week if we so desired. A couple of miles away, over the hill, was the American base, and we had a happy exchange of social activities: it took a pretty good storm to stop us from having our weekly bridge game. Commander Flynn was a very shrewd player, and although we tried very hard we never succeeded in defeating him.

The garage was perhaps the busiest place on the whole base. Day after day, we overhauled the vehicles and modified them for the journey south. We improved their suspension, built canvas cabs to give some protection against the bitter wind, and removed any equipment that might be adding unnecessary weight. As we had no heating in the tractors, I was determined that we must have a place where we could get protection and warmth. So we set to work to build ourselves a small van on skis, which we called "the caboose." In it we had bunks for two men, our communications radio, a cooking bench, and a hot-air blower. There was nothing fancy about the caboose, but on the trip south it proved a lifesaver.

It was tremendously exciting when the sky to the north started to lighten and we knew that the sun was returning. The tempo of life increased—there was still so much to do.

When the sun peeped back over the horizon, dogs and men rushed out across the bay ice toward it. It is an unbelievable experience to see the sun again after the long, dark winter. Scott Base was now a secure and well-proven home, capable of withstanding any storm and anchored down by drift snow.

Our thoughts turned increasingly toward field activities. In the cold, dim light of spring, we did many long trips of exploration with dogs and vehicles—testing out equipment, laying depots, and preparing ourselves for our summer journeys. Our Ferguson tractors were now ready for action. They were incredibly primitive-looking contraptions, but they had shown that they could handle firm snow with reasonable success. Most important of all, the engines seemed to keep going whatever happened. Only deep, soft snow and crevasse areas remained major hazards.

On October 14, 1957, we started on our southward journey. We were given a warm farewell—but those left behind had little confidence in our success. I am sure that few of them (including Admiral George Dufek) thought we would get more than fifty miles out on the Ross Ice Shelf with our strange vehicles. We were still in sight of the cross on Observation Hill when we struck our first problem. A heavily laden sledge broke through the top of a concealed crevasse, and we had to unload the drums of fuel in order to extricate it. We pushed on again, only to strike soft snow, which caused the greatest difficulty in moving our loads. That night, we camped only a few miles from Scott Base, and I wondered if the forecasts of failure would indeed prove to be true.

Next day, we reduced our loads, desperate to make more mileage, and we began to move faster and faster. For day after day we made our way across the great ice shelf. Whenever the surface was hard, we made good time.

When the surface was soft, we labored and struggled. It was a tremendous moment when we reached the bottom of the Skelton Glacier and pulled up alongside the depot we had established the previous autumn.

Already, we were experiencing mechanical problems, and our Weasel required replacement of a part. Our two ingenious engineers, Murray Ellis and Jim Bates, rigged up a bipod, lifted the engine out of its housing, replaced the part, and then lowered it all back into place—everything done in 30°-below temperatures.

I regarded the Skelton Glacier as the most unpredictable part of our journey. Would we be able to make the nine-thousand-foot rise in altitude that it entailed? What problems would we have with the many crevasses that could be seen so clearly from the air? In the lower Skelton we were constantly harassed by wind sweeping down from the snowfields above. Visibility was minimal and we just had to push on, hoping we could keep clear of the worst crevasse areas. We commenced bumping over row after row of crevasses, although few of them at this stage were large enough to cause us much distress. As we started gaining height among the peaks, the winds decreased, but our fear of crevasses became much greater. We were having to wind our way in and out through unpleasant open areas, and there were few moments of the day in which we weren't operating in constant tension. We had no crevasse-detection equipment apart from the tractors themselves, and frequently our first indication that we were in a dangerous area was when the back wheels would knock a hole in the lid of a crevasse.

The upper Skelton Glacier was a beautiful place, and for a while the surface was much easier as we traveled under the great slopes of Mount Huggins. We reached the névé of the glacier and became involved in constant wind and drifting snow. Despite our fear of crevasse areas, we pushed on, only hoping that luck would be on our side. Navigation became a considerable problem, and frequently visibility became so bad that we had to stop and wait for a few hours until the winds had eased. Temperatures were mostly around 30–40 below, and it was unpleasant work driving in our unheated cabs. It was a tremendous moment —for me perhaps the greatest on the expedition—when we emerged through the driving snow to see the site of Plateau Depot half a mile away. Despite all forecasts, we had climbed the Skelton Glacier and reached the open spaces of the Antarctic Plateau. Bad weather at the Plateau Depot severely restricted the flying in of supplies, but slowly the job was completed and we were able to load the sledges and move on again.

We traveled west and south in an area of constant winds and drifting snow. Some of the surface was very hard indeed, with giant sastruga that tipped over many a sledge. And then we struck area after area of crevasses, and frequently our first notification was when the tractor would sink down into one. The vehicles were roped together for protection, but we knew this would be of limited value in a really big crevasse. Our tracks became peppered with open holes—and I never really got quite used them. Depot 480 was established 480 miles from Scott Base, and at 81°30″ south we supplied Midway Depot.

Life had become a routine of cold, bad visibility, and crevasses. We had many close shaves with disaster, and it was a miracle we didn't lose any vehicles. We became very familiar with the insides of crevasses—an experience most of us could well have done without.

We established Depot 700 on December 15, 1957. This was our final depot for the crossing party, and it was already much farther out than had originally been planned.

We were now five hundred miles from the Pole, and an awfully long way from anywhere else. Fuchs was making very slow progress; he had considerable problems with severe crevasse areas, and he was methodically carrying out a detailed seismic program. Already, the media were talking about "a race for the Pole," but this was a considerable exaggeration. It was unfortunately true that Fuchs and I didn't communicate terribly well—he was reluctant to commit himself too freely—while I liked to know what was going on. If I didn't hear anything, I just made the decisions myself. I suppose what Fuchs really needed was a good, reliable navy officer who would do the job he was chosen to do.

At Depot 700 we set up our radio mast and installed our aircraft homing device. After a magnificent flight across the Antarctic glaciers, our Beaver aircraft landed beside the tractors and we unloaded its mail, fuel, and supplies. For five days, our pilots operated at the absolute limit of the aircraft range and flew in more drums of fuel. Finally the depot was completed and there was fuel left over, too—twenty drums of it. It was enough, I felt, to get us to the Pole, but no farther.

On December 20, we headed south again with our three Ferguson tractors (our Weasel had succumbed a hundred miles back). During the first fifty miles, we struck a series of crevasse areas—some of them bad crevasses. We had to reconnoiter ahead on foot, testing the ground with ice axes, estimating the safety of flimsy snow bridges. It was a difficult and nerve-wracking sector, but things soon started to improve. The crevasse areas became less frequent and we were able to make better daily times.

A couple of hundred miles from the Pole, we struck one of our worst obstacles—soft, bottomless snow. We tried everything—relaying our loads, various combinations of vehicles—but it was hopeless. We couldn't move. We established a depot and got rid of every bit of excess equipment and sledges. We carried on with a minimum of gear, knowing that we were cutting things very fine. With our lighter loads, we could just move, but fuel was being consumed at a terrifying rate. I wondered whether we would be forced to abandon a vehicle. We decided to do without sleep and just keep driving. On January 3, our tired eyes picked up a tiny dot ahead of us; it was a flag, placed there by men from the American Pole Station. We'd made it. Next morning, January 4, we reached the Pole Station, the first men to drive vehicles overland to the South Pole.

I had found our Antarctic expedition a very demanding, but satisfying, experience. My party had carried out an extensive scientific program and had explored and geologized over thousands of square miles of new terrain. With dogs and tractors and aircraft, we had laid five depots for Fuchs's crossing party and reconnoitered the route. The trip to the Pole had been an exciting extra; it may not have added much to the cause of science, but it had shown that if you were keen and resourceful enough, you could get even a bunch of farm tractors to the South Pole—and in these days of increasingly expensive equipment, that must be something.

Two weeks later, Fuchs and his party approached the Pole from the other side of the continent, their powerful vehicles dragging their loads with remarkable ease. After a few days at the Pole, they carried back along our tracks. I joined them at Depot 700 and helped guide them across the plateau and down the Skelton Glacier. On March 2, we reached Scott Base; the journey had been completed, the Antarctic Continent crossed. Shackleton's dream had become reality.

It was nine years before I returned to the Antarctic, and

great changes had taken place: there had been widespread tractor journeys and sophisticated scientific programs. Instead of battling our way south in an old wooden vessel, we flew in a civilian airliner and landed smoothly on the ice in McMurdo Sound. Our program was neat and compact—geology and mountaineering in the Cape Hallett area, and a journey north to Robertson Bay. We had a couple of lively snowmobiles and four heavily laden sledges. We drove to the skiplane field on the Ross Ice Shelf and loaded our vehicles and sledges into a Hercules—the perfect aircraft for Antarctic conditions. With Admiral Abbot at the controls, we flew north over the clouds and above the pack ice. The clouds rolled back to show us the mighty Tucker Glacier, and before long we were landing on the rough sea ice near Hallett Station. Across the bay, we could see the magnificent spire of Mount Herschel, our major climbing objective. Cape Hallett is the site of an enormous penguin rookery, and the sound and the smell were something to experience. We headed across the bay ice toward Mount Herschel. At first the traveling was quite easy, but after a while the snow softened and we had the greatest of difficulty in moving our heavy loads. We passed under mighty icebergs—ominous indications that we were living on an unstable surface that sometime in the future would all go out to sea. We learned to handle our snowmobiles even in the deep snow—riding them like bucking broncos, staying on course and compacting a trail—but we made many mistakes until we learned the correct technique. After a few trips, the trail would be firm enough to handle a heavy sledge.

The bay ice was full of tide cracks and seal holes. As you lay in your tent, you could hear the seals breathing noisily underneath your bed, and it all felt a little insecure. We struck up the Maubray Glacier toward the coast. As a protection against the multitude of crevasses, I strapped a plank of wood across the front of my snowmobile. A blizzard forced us to camp, and for four days we were battered by as strong winds as I have experienced. The tents had been built to my own design and had never been tested in this sort of weather before. I didn't sleep much the first night, but gradually my confidence grew as the tents withstood the storm.

We had other problems, too. Murray Ellis and I were traveling, through poor visibility, with our snowmobiles roped together. Suddenly the lid of a crevasse collapsed under his vehicle. He flicked the throttle full open and teetered on the brink, then just managed to clamber up the edge onto safe ground. Following behind, I only had time to shut off the throttle and turn my vehicle to the side. Between us was a gaping vent, dropping away several hundred feet. It took a few minutes before our nerves calmed sufficiently so that we could extricate ourselves from this unpleasant position.

In strong winds, we battled our way up onto the coast line. From five thousand feet, we looked over Robertson Bay, and in the distance could see Cape Adare, where Borchgrevink first wintered, in 1899. His hut still stands there today. We carried out an extensive geological survey program and then retreated back to the bay ice and turned our attention to Mount Herschel. Mountaineering in the Antarctic has its own problems. The very low temperatures and the isolation mean that the party must always carry warm clothing and protection. To climb Mount Herschel, which is nearly eleven thousand feet high, we had to start right from sea level. Base camp was pitched among the tide cracks, and we had frequent visits from inquisitive Weddell seals. The bay was surrounded by a vertical ice wall with no obvious line to climb it by. We searched the bottom of the ice wall, seeking a safe approach. Often, our legs disappeared through tide cracks into the frigid ocean. Then we found a way up—a contorted and difficult one, leading in and out of crevasses and over ice pinnacles. When we reached the top of the ice wall we were still only a couple of hundred feet above sea level, and it had taken us three days. Carrying heavy loads of supplies, we battled our way up a steep, crevassed valley. It was hard going, as the snow was very soft, and we perspired freely from the effort. The perspiration seeped through our woolen sweaters and then froze in knobbly lumps on the outside. At the head of the valley it became very steep indeed, and with considerable care we climbed the last slope, hacked a hole through the cornice, and emerged on a saddle on the crest of the ridge. It was a broad and roomy place, ideal for a camp. We pitched a tent there, saw two members of the party well established, and then returned down the long valley to our camp on the sea ice.

Next day, Dr. Michael Gill and Bruce Jenkinson made a mighty push for the summit. They had six thousand feet to go, much of it over steep and difficult terrain. From their camp, they climbed an exposed slope onto the crest of the ridge; this took quite a time. Then a great ice band barred their way—a much tougher problem. Hacking steps with their crampons, they made their way up this steep surface, wriggling in and out of crevasses and surmounting ice walls. It is doubtful if climbing of this standard had ever before been carried out in the Antarctic. With the ice wall behind them, they had an enormous slope to ascend—over a thousand feet long and extremely demanding on leg muscles and balance. With a great feat of agility and strength, they forced their way up the slope and emerged on the north shoulder. Ahead of them now was a long ridge leading toward the sharp summit cone. The ridge itself was long and demanding and took a long time, but finally they reached the foot of the summit pyramid. This was steep and got even steeper as they progressed up it. They were protected from the wind here and the work was hard, so they perspired freely. The angle got steeper and steeper, and they wondered if they should persist. With determination, they pushed on and finally hauled themselves up the last few feet to emerge on top of the mountain. They had made the first ascent of this magnificent Antarctic peak.

Their descent of the mountain became a classic of endurance. It was late and very cold, and they had been going for nineteen hours. Far across the bay they could see Cape Hallett, where the trip had started and where it would end again in a few days. They could see, too, signs of the ice breaking out, and they knew that the trip must end soon. It had been a good trip and a successful one. With modern equipment and communications, a lot had been done in a few short weeks. It seemed hard to believe that in a couple of days we would all be back on the green grass at Christchurch Airport, in New Zealand.

III

What is the future of the South Pole area? It is still a continent of adventure: there is still pack ice to be traversed, mountains to be climbed, and valleys to be explored. The coast line is very beautiful and there is a surprising amount of marine and bird life. The Antarctic has seen much international cooperation and little discord. It is an example to the world. I hope it stays that way.

We hear so many rumors of Antarctic oil discoveries and mineral wealth, suggestions about Antarctic grain storage—if there is any grain to store—and atomic waste disposal. Heaven protect the Antarctic from commercial exploitation, which will almost surely pollute and destroy the continent.

The Antarctic has much to offer our young people: challenges to be overcome, scientific puzzles to unravel, distances to be covered, dangers to be met. It is a continent where the line between success and failure can be very fine; where courage and dogged determination must be paramount; where comradeship is more important than personal glory. Let us hope it remains that way—a reservoir of adventure, a challenge to the spirit of man.

ON THE ORIGIN
OF THE UNIVERSE

SIR FRED HOYLE *British astronomer and mathe-
matician Sir Fred Hoyle divides his energies between Eng-
land and the United States, spending part of each year at
California's Hale Observatories and in teaching posts at the
California Institute of Technology and Cornell University,
and the remainder researching and writing at his home in
Cumberland, England. For many years, Hoyle was Plumian
Professor of Astronomy and Experimental Philosophy at
Cambridge University in England, and Director of the
Cambridge Institute of Theoretical Astronomy.*

*Hoyle is a leading proponent of the "steady-state"
theory of universal creation, a theory of "continuous creation" first proposed with Thomas Gold and Hermann
Bondi in 1948. Since 1948, the "steady-state" theory has
been modified to account for oscillations in finite regions of
the infinite universe. Controversy continues between
"steady-state" theorists and advocates of a "big bang"
theory of creation. Hoyle is also recognized for significant
contributions to the study of stars and quasars, their ori-
gins, ages, and structures.*

*In addition to numerous scientific works on astronomy
and popular science, among them* THE NATURE OF THE UNI-
VERSE *(1950) and* FRONTIERS OF ASTRONOMY *(1955), Hoyle
has written many science-fiction novels, a play, and the
libretto for an opera.*

I

ASTRONOMERS BELIEVE that some 15 billion years ago the universe was in a state very different from the condition observed today. Hitherto, it has not been possible to describe the situation existing before this particular time, which has therefore become known as the "origin" of the universe. In the first part of this lecture we shall consider this usual view of cosmology.

The main substance of the lecture lies, however, in the development of a state for the universe existing *before* this so-called moment of origin. The point of view described here is that the previous failure to extend time backward beyond the so-called origin arose from a wrong technique for handling the problem, rather than from any actual physical limitation on the universe itself.

It is, then, of interest to inquire into the state of affairs before the "origin" particularly with reference to whether the preceding state of affairs has had an influence on the present-day situation. There are many aspects of astronomy that seem to imply such an influence, as if information coming from an earlier epoch has been impressed on the observed distribution of galaxies and stars. We shall find that a condition for the universe preceding the "origin" can very well be the source of this information.

II

Figure 1 shows the way in which physicists measure both time intervals and spatial distances. The wavelength of monochromatic radiation determines a unit both of time and of spatial distance, in the manner indicated. The single spatial distance x represents a shorthand for the actual three dimensions of space. With this understanding, figure 1 is said to be a spacetime diagram.

This procedure is an essentially local one, suitable for the physicist concerned with events occurring in a terrestrial laboratory but not applicable over big distances or over a long span of time—not suitable for the astronomer concerned with the structure of the universe in the large. To cover the whole universe, we need many observers, situated in different localities, with each of the observers using the method of figure 1 within his particular locality. But this raises the problem of how we are to collate the different observers one with another, how we are to fit together many pieces in order to obtain a picture of the whole universe.

The general method of attack on this problem was formulated in the nineteenth century by G. F. B. Riemann. The Riemann method is complex and capable of giving rise to a bewildering variety of pictures for the universe. However, the astronomer avoids the worst complexities by insisting that the resulting picture satisfy two important restrictions. One of these restrictions is strongly supported by actual observations of distant galaxies, whereas the other restriction fits an intellectual concept dating from the time of Copernicus.

Copernicus changed the earlier view of the Earth as the

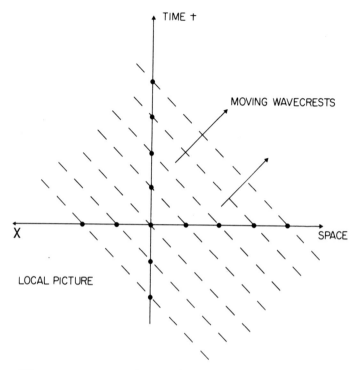

1. The wave structure of monochromatic radiation can be used to determine units both of time and of spatial distance.

center of our solar system to a picture in which the Earth has no such privileged position. Later generations have extended this idea to a situation in which our solar system has no privileged position with respect to the galaxy in which we live, and in which our galaxy has no privileged position with respect to other galaxies. As viewed from any galaxy, the universe is regarded as looking the same. This concept is known as the principle of *homogeneity*.

In addition to the principle of homogeneity, astronomers also suppose the universe to have no directional structure:

On the Origin of the Universe

the large-scale features are to be the same in one direction as in another direction. This principle of *isotropy* is attested to by observation, for we do find the large-scale properties of the universe to be the same in all directions.

Given these two principles, of homogeneity and of isotropy, it becomes possible to fit together the spacetime diagrams of observers in different galaxies, and to do so in a remarkably simple way. Thus the time measurements of all such observers, each using the method of figure 1, turn out to be synchronous: they are all the same, as is illustrated in figure 2. Subject to the principles of homogeneity and isotropy, such measurements are said to determine a system of *universal time*. We shall consider time in this sense throughout the present lecture.

What can we now say about the motions of the galaxies? Each galaxy can be considered as a tube in our spacetime diagram, and we can approximate each such tube by a line, as in figure 3 for a particular galaxy. At each moment of the time, such a line is characterized by a spatial point. Suppose we consider the spatial points determined by a particular moment of time for a set of n galaxies, and let us call the corresponding points G_1, G_2, G_3, . . . G_n. Then we can form a spatial polygon by joining G_1 to G_2, G_2 to G_3, . . . G_n-_1 to G_n, G_n to G_1. This procedure can be repeated for another moment of time. What can we say about the two resulting polygons?

Consistent with our two principles of homogeneity and isotropy, we have the situation of figure 4, drawn for the case of $n=5$. *The two polygons must have the same shape, but they can be of different scale.* The same property holds good whatever value we choose for n.

Equipped only with the two principles of homogeneity and isotropy, we have no means of knowing the time sequence of figure 4. If we wish to know which of these

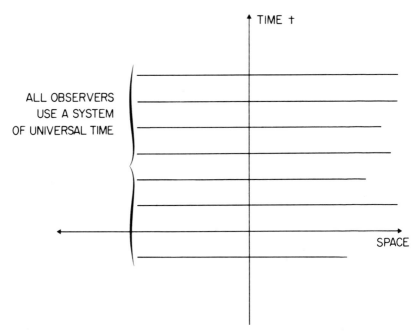

2. The principles of homogeneity and isotropy lead to a system of universal time.

two polygons occurs later in our spacetime diagram, we must appeal to observation. We have the situation that if the larger polygon of figure 4 occurs later, then the galaxies are expanding apart from each other. In the contrary case, the galaxies would be approaching each other as time went on. The situation determined by astronomical observations is that the galaxies expand apart from each other, so that it is the larger polygon of figure 4 that occurs later in time.

It is useful to characterize the scale of our galaxy polygons by a factor that we will call Q, with Q changing as the time t changes. This behavior we denote by using $Q(t)$. Furthermore, we can choose some particular mo-

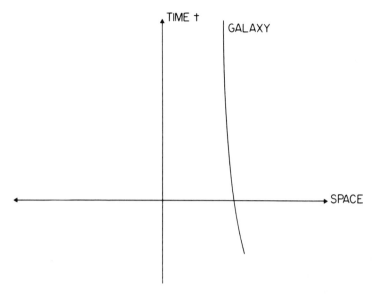

3. A galaxy is a tube in four-dimensional spacetime, the tube being approximated here as a line.

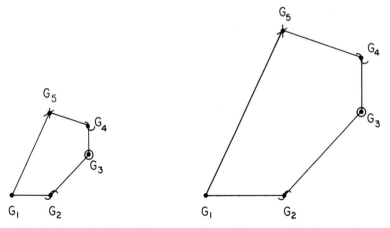

4. Under the principles of homogeneity and isotropy the shapes of two polygons, obtained by joining the spatial positions of a set of galaxies at two different moments of universal time, must be the same. But these principles do not require the two polygons to have the same scale. Nor do these principles determine which of the two polygons occurs later in time.

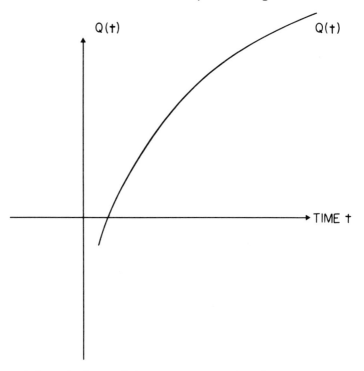

5. Schematic form of the curve representing the behavior of the scale factor $Q(t)$ with respect to the time t.

ment, say t_0, as a standard of reference by setting $Q(t_0)=1$. This means that we take the polygon for time t_0 as our standard, and the scale of the polygon at other times is then reckoned with respect to the scale of time t_0.

If the only important force operating between the galaxies is that of gravitation, we can say that $Q(t)$ must behave in the manner of figure 5. That is to say, the curve of Q plotted against the time t must be such that the tangent always turns in a clockwise sense. This is because gravitation is an attractive force; a force of repulsion

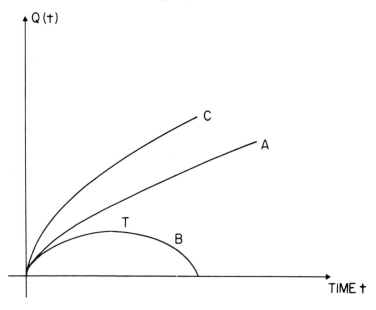

6. The behavior of $Q(t)$, according to Einstein's gravitational theory, for the cases A, B, and C.

would cause the curve of Q to turn in the opposite, anticlockwise sense.

By using the precise theory of gravitation formulated by Albert Einstein, it is possible to go somewhat further than this, but not far enough to determine a definite form for the curve of Q. What we can say, however, is that the curve must be of one of three types, A, B, or C, shown in figure 6. Curve A has a precise form, separating curves of type B from those of type C. A curve of type B has the property that gravitation is strong enough eventually to stop the expansion apart of the galaxies. Thus, expansion ceases at the time when Q attains the maximum of the curve B of figure 6. Thereafter, the expansion is replaced by contraction. For curves A and C, on the other hand, expansion never ceases. Eventually, for large enough t, gravitation weakens to a stage where the curve of Q becomes a straight line, as can be seen in figure 6.

It is an outstanding ambition of astronomers to determine which kind of a Q curve the universe actually follows. A few years ago, many astronomers believed that a type B curve was correct, but today there are many who think type C to be correct. In what follows, I shall adopt the intermediate A, although nothing that will be said depends crucially on this step; we take it simply for definiteness in the discussion.

It is a common property of all these possibilities that the scale factor Q was zero at a definite moment in the past, as is clearly shown by figure 6. We can see what this means by returning to our galaxy polygons. In figure 7 we have the case of a set of three galaxies, the polygon being a triangle. In the future, the triangle for the three galaxies will be larger than it is at present. In the past, the triangle was smaller, and far enough back into the past the triangle was

PAST   PRESENT   FUTURE

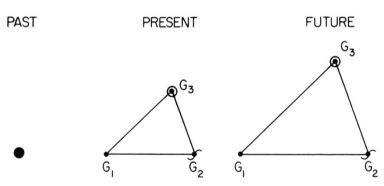

7. The triangle formed by joining the positions of the three galaxies will be larger in the future than it is at present. In the past the triangle was smaller, and far enough back into the past the triangle was shrunk down to nothing at all.

shrunk down to nothing at all! This was the moment at which Q was zero. It is this moment, believed by astronomers to have been between twelve and fifteen billion years ago, that is frequently referred to as the "origin" of the universe.

Do we have evidence that such a situation, with the galaxies all pressed together, ever really existed? There are no observations relating directly to the galaxies themselves to confirm that this was so. Thus, observations of galaxies extend backward in time for about 5 billion years, about one third of the way back to Q=0.

Now, although galaxies are not observed back in time sufficiently to confirm the occurrence of the peculiar condition Q=0, radiation known as the "microwave background" is indeed observed and is believed to have been generated at a time very close to the origin of the universe, as is shown in figure 8. This microwave radiation is mostly observed over the wavelength band from about one millimeter to about ten centimeters, just the band in which radi-

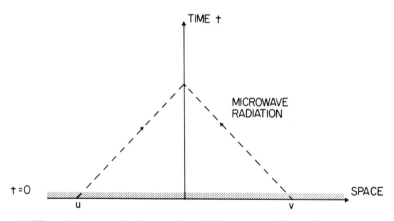

8. The microwave background radiation comes from the remote past, being generated at a time close to $t=0$.

ation is observed from the discrete radio sources. Since the radio sources are mostly galaxies, we can ask how we can be sure that this microwave radiation really does come from the early universe and not from galaxies at much later times. Two answers are given to this question. First, the known discrete radio sources do not produce a sufficient intensity, especially at the shorter wavelengths, to explain the observations. Second, the observed background is distributed more smoothly over the sky than would be the case if the radiation were generated by discrete sources of the type we know about. So, on both these grounds it is thought that the microwave background really does confirm the existence of the moment $t=0$, when the scale factor Q was zero. As the term "origin" of the universe implies, this moment is usually taken to represent the creation of the universe, in the sense that the universe existed after $t=0$ but not before $t=0$.

This concept raises a peculiar difficulty. Microwave radiation is essentially the same no matter from what direction it reaches us. It is the same from the two distinct directions shown in figure 8, which requires the physical conditions near the points U and V to be the same. But, according to the usual view, the regions near U and V were never in communication with each other. So how could they "know" to be the same? Within the usual view, this question and others of a similar nature seem unanswerable.

In this lecture, I shall seek to supply an answer to this problem. Referring back to figure 7, I shall try to penetrate back through the moment $t=0$, when it seems as though the material of the galaxies all came together. What we shall do is to build an intellectual microscope to enable us to see through this moment, to conditions that lie beyond it. In other words, we reject the concept that the universe did not exist before $t=0$. We seek to discover what lay before $t=0$, and we do so with the aim of solving certain explicit problems. First, how did the microwave radiation come to be present near $t=0$? Second, why were the conditions near U and V of figure 8 so similar? And third, why is the energy of the microwave radiation observed to be so similar to that of the light generated by the stars of the galaxies?

When we say the galaxies expand apart from each other as time goes on, we mean that the spatial distances between them increase relative to a scale set by the method of figure 1. Now, the radiation in this figure is to be regarded as generated by some standard laboratory procedure. In prac-

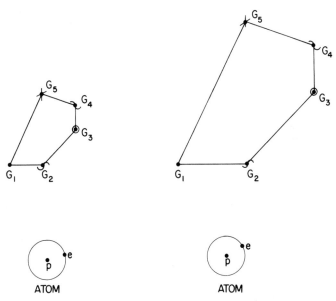

9. When we say that the scale of the triangle formed by any three galaxies increases with time, we mean that the *ratio* of the triangle to the spatial length unit determined by atoms (cf. figure 1) is increasing.

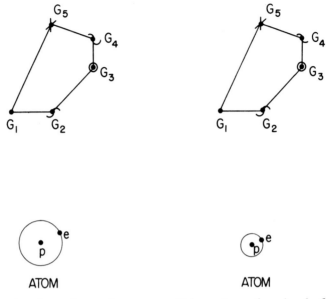

10. An alternative to figure 9 would be to keep the triangle fixed and to suppose that the atomic dimensions increase with time.

tice, radiation from a certain transition of a particular atom is used. So the basic spatial scale, relative to which the galaxies expand apart, is set by atoms in the laboratory. In fact, we have the situation of figure 9, with the scale set by the atoms being regarded as always the same—i.e., not changing with time. But could we replace figure 9 by figure 10, with the polygon formed by the galaxies staying the same and with the atoms changing? Let us see how we might seek to give expression to this idea.

The lines followed by the galaxies now become parallel to each other in the spacetime diagram, as in figure 11. Next, imagine the material of the galaxies to be smoothed out into a uniform distribution, as in figure 12. The average

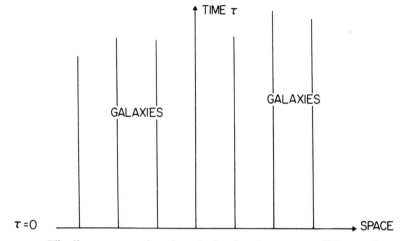

11. The lines representing the galaxies then become parallel to each other, and to the time axis.

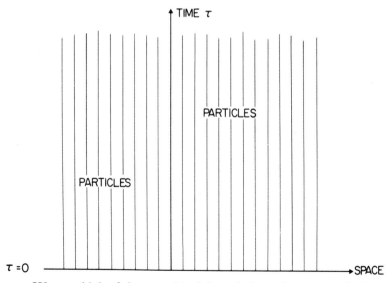

12. We can think of the material of the galaxies as being smoothed out into a uniform distribution of particles.

distance apart of the particles in this smoothed diagram gives us a new spatial scale, which we use now instead of the scale set by the procedure of figure 1. We also change the scale by which time intervals are to be measured, simply by arranging for radiation to propagate at an angle of 45° in our new spacetime diagram; this sets the same scale for time intervals as for spatial distances, a procedure illus-

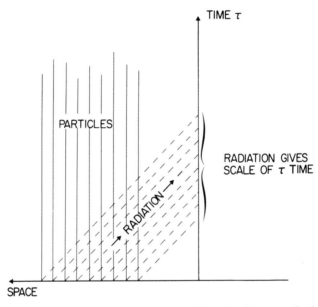

13. The average distance between particles in this smoothed-out picture gives a spatial scale. The scale in which time τ is measured is chosen so that radiation propagates at an angle of 45° to the time axis.

trated in figure 13. We refer to time measured in this way by τ, to distinguish from the time t determined by the method of figure 1.

The situation that now emerges is this: provided the masses of all particles are considered to vary with time proportionally to τ^2, nothing is changed when we pass from the picture of figure 9 to that of figure 10. Every observable quantity connected with the large-scale structure of the universe remains unchanged. Hence, we may just as well use figure 10 together with figure 13 and with particle mass proportional to τ^2, as use figure 9. And now we are no longer faced with the situation of figure 7, and with the

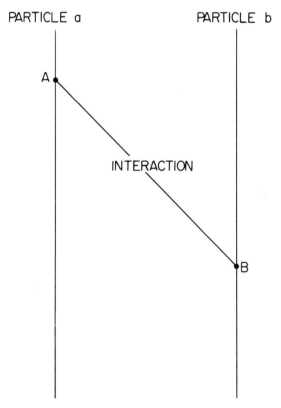

14. The mass of particle *a* at the point A is to be regarded as made up of contributions from other particles *b*, traveling also at an angle of 45° to the time axis.

conceptual difficulty of the scale factor Q being zero. In short, we have constructed our intellectual microscope.

It emerges, further, that we can readily understand why the particle masses behave like τ^2. For calculation shows that, if the mass of a particle arises from interactions with other particles, a behavior similar to that which is well known for an electrical field leads to the proportionality τ^2. Such interactions are illustrated in figure 14.

As an aside, it is worth noticing how the so-called red shift of the light from a distant galaxy receives a particularly simple explanation in this way of looking at things. In figure 15, the light from a galaxy at distance r is consid-

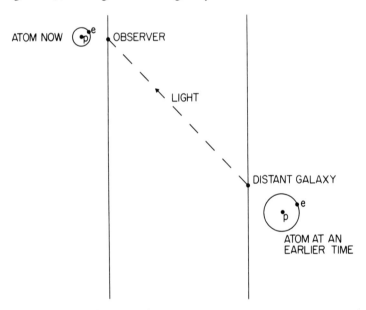

15. We can say that light from a distant galaxy was generated at a time when the masses of particles were less than they are at present, and it is for this reason that radiation from an explicit atomic transition is red-shifted with respect to present-day radiation from the same transition.

ered to be received by us at time τ. The light from the galaxy is required to have started its journey at time $\tau - r$, when the masses of the emitting atoms were less than the masses of present-day atoms by the ratio $(\tau - r)^2/\tau^2$. Because the particle masses were less, the frequency of the radiation emitted in the transitions of a particular atom was less than the frequency of similar radiation from the transitions of present-day atoms, again the ratio $(\tau - r)^2/\tau^2$. It is just this ratio which leads to the well-known relation between the change of frequency of radiation and the distance r of the emitting galaxy. All the usual cosmological results can be obtained in this way.

At this point we have to ask whether the structure of the universe can really be as crude as it is shown to be in figure

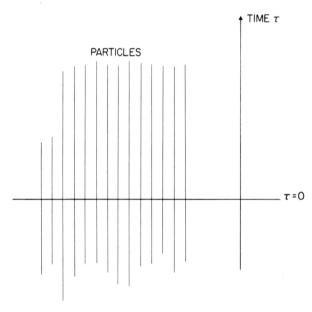

16. We can seek to avoid breaks in the paths of particles by extending them indefinitely backward in time.

12. Do the particles all begin their existence precisely at the moment $\tau=0$? Let us attempt to answer this question by considering what would happen if the lines of the particles were extended backward before $\tau=0$, in the manner of figure 16. Now compute once more the particle masses, following the method of figure 14. In this present situation, contributions to the mass of a particle come from interactions with other particles going back farther and farther into the past: the interactions are no longer cut off at $\tau=0$. Then, instead of the particle masses being proportional to τ^2, it turns out that masses are infinite, a wholly unacceptable result. Hence, extending the lines of the particles backward in the manner of figure 16 seems at first sight to fail. Indeed, it seems as if it is just because all the particles are considered in figure 14 to begin at $\tau=0$ that we obtained the required proportionality τ^2 for the masses. Yet, if we proceed as in figure 17 the situation is suitably restored. Here the particles extend backward from $\tau=0$, but at times before $\tau=0$ they contribute negatively to the masses. Provided the mass interaction switches sign at $\tau=0$, all the previous results continue to hold good. We are not obliged, therefore, to require the whole universe to have begun at $\tau=0$.

Notice the important point that *two signs* are involved in determining the sign of a mass contribution. A sign is involved at each end of the interaction. If each end is at a time later than $\tau=0$, both signs are positive, and their product is positive. If each end is at a time earlier than $\tau=0$, both signs are negative *and their product is again positive*. Thus mass contributions are positive whenever both ends of an interaction lie on the same side of $\tau=0$. But when the two ends of an interaction lie on opposite sides of $\tau=0$, mass contributions are negative.

This means that particle masses behave symmetrically

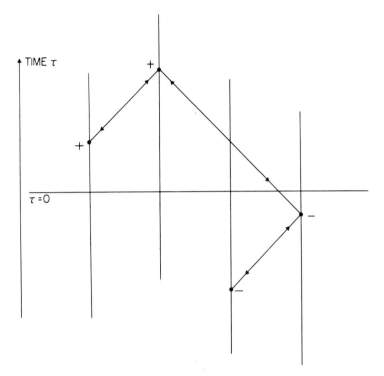

17. Illustrating the hypothesis that positive and negative mass interactions are separated at $\tau=0$.

with respect to $\tau=0$, just as the proportionality τ^2 implies $[\tau^2=(-\tau)^2]$. The situation before $\tau=0$ is a kind of mirror image of the situation after $\tau=0$, as can be seen from the plot of the mass of particle against time τ shown in figure 18.

We are now close to being able to clear up a number of apparently mysterious problems. One important further idea is still needed, however. Radiation propagates only one way with respect to time. It was already implied in figure 1 that radiation propagates in the sense in which the time

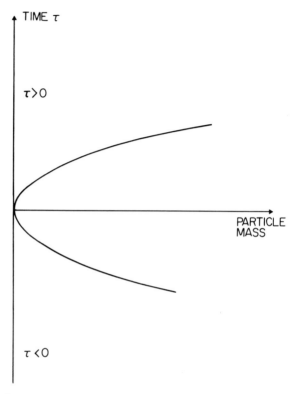

18. The mass of a particle is proportional to τ^2, which gives a similar behavior for times before and after $\tau=0$.

t increases, and a similar supposition applied to figure 13. This property of radiation accords with the world of experience *as it applies on our side of the moment $\tau=0$.* But what of the other side of $\tau=0$? How does radiation propagate at times earlier than $\tau=0$? We answer this question in the manner of figure 19, with radiation propagating *in the same sense* both before $\tau=0$ and after $\tau=0$. Otherwise we would be faced with a switch (i.e., a discontinuity) in the sense of propagation, occurring at the moment $\tau=0$.

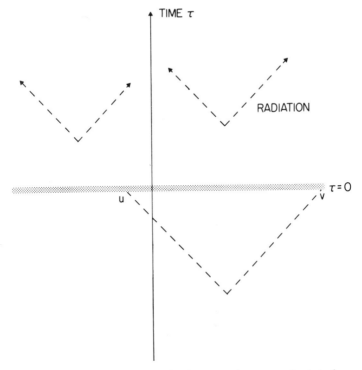

19. Radiation progagates always in the same time sense, both before and after $\tau = 0$.

With the situation of figure 19, the two sides of $\tau = 0$ are not physically the same, since on the "other side" radiation travels toward $\tau = 0$ while on "our side" radiation travels away from $\tau = 0$. Instead of the red-shift effect of figure 15, an observer on the other side finds an opposite, blue-shift effect. Thus the frequency of light from a distant galaxy, emitted in the particular transition of a certain atom, is greater than the corresponding frequency of similar atoms in his own locality, the ratio being again

$(\tau-r)^2/\tau^2$ but with τ now a negative number, and so with this ratio the frequency is greater than unity.

This is on the assumption that stars and galaxies exist on the other side, just as they do on our side. Starlight generated on the other side propagates toward $\tau=0$, in the manner shown in figure 19. What happens to this starlight? Can it come right through $\tau=0$, and so permit us to observe the galaxies on the other side? The answer to this question is, No. Such radiation is powerfully absorbed near $\tau=0$, because the particles of gas lying between the stars have individual masses that are very small near $\tau=0$ and particles of small mass are exceedingly powerful absorbers of radiation, very much more so than the particles of our present-day environment. Such particles are also powerful emitters of radiation, and they immediately re-emit the absorbed radiation, so that, in a sense, the radiation does indeed come through the moment $\tau=0$. However, the galaxies on the other side are entirely blurred out by this absorption and re-emission process, and so cannot be observed at all in the usual way.

Nevertheless, the blurred radiation crossing through $\tau=0$ from the other side *can be observed, and it is observed as the microwave background.*

Let us return now to the situation of the points U and V of figure 8. It will be recalled that the microwave radiation from two such points is observed to be very similar, which creates a difficulty for the usual theories of cosmology, because in the usual theories these points have never been connected in any way. Now, however, we see that conditions at U and V are similar because these conditions have been generated by a spatially homogeneous situation on the other side of $\tau=0$. The points U and V can even receive radiation from the same stars and galaxies.

A third point is that the energy carried by the microwave background is that of the starlight generated on the other side. Since it is reasonable to suppose that this latter energy is of a similar general order to the energy of starlight generated on our side, we also explain the general coincidence between the microwave energy and that of starlight on our side. Hence, both the energy content and the uniformity of the microwave radiation follow immediately from its mode of origin.[1]

We are well used to the idea that radiation is capable of carrying information; this forms the basis of the use of radio transmissions as a means of communication. There is a flow of information in the time sense of the propagation of the radiation, but not in the opposite time sense. Thus, on "our side" of $\tau=0$ the flow of information is directed *away* from $\tau=0$, whereas on the "other side" the flow is *toward* $\tau=0$. It is relevant, then, to ask if the flow on the other side actually crosses $\tau=0$ and so passes information to our side. An affirmative answer to this question would imply that aspects of our astronomical observations could depend on the physical conditions occurring on the other side. Indeed, in such a case, we might, from astronomical observations, be able to infer what the conditions on the other side actually were. This might become a prime aim of astronomical studies.

The same idea can be thought of in terms of cause and effect. The normal time sequence of cause and effect, from past to future, is the same as the time sense for the propagation of radiation. In these terms, we can ask how far the events observed by astronomers are "caused" by the situation that existed at times earlier than $\tau=0$. An affirmative answer to this question would imply that astronomical theories must remain substantially incomplete unless they contain a discussion of the "other side" of $\tau=0$.

Information is certainly lost at $\tau=0$. We have already

seen that galaxies on the other side could not be observed directly, because starlight on the other side becomes blurred into the smooth microwave background. On the other hand, stars existing before $\tau=0$ may well persist through $\tau=0$, emerging as stars on our side. It is true that the ability of a star to hold itself together by gravitational attraction becomes exceedingly weak as the masses of its constituent particles become very small. The particles of a star then become susceptible to evaporation, by the very radiation field that the stars themselves created at times appreciably before $\tau=0$. Yet, calculation suggests that any such evaporation can be only partial; a stellar condensation is maintained through $\tau=0$. Thus, showers of stars come through $\tau=0$, from the other side to our side.

It also seems likely that the structural forms that we call "galaxies" may be the result of information flowing to our side across $\tau=0$. The galaxies may well be memories of structural forms that existed on the other side.

The last question that I wish to consider in this lecture is the nature of the particular moment $\tau=0$. In the interpretation developed above, it is at this moment that mass interactions switch their signs. How does it come about that the switches all happen to occur at a particular moment?

Consider the situation of figure 20, in which the $+$ and $-$ aggregates have quite general shapes. At particles in a positive $(+)$ aggregate, the mass interaction has a plus sign, while at particles in a negative $(-)$ aggregate, the mass interaction has a minus sign. Once again, each mass interaction—taking place between one particle and another—has two ends, and we must take the product of the appropriate signs at the two ends. If both ends are in a positive aggregate or both in a negative aggregate, the product is positive.[2] But if one end is in a positive aggregate and the other end in a negative aggregate, the product is negative.

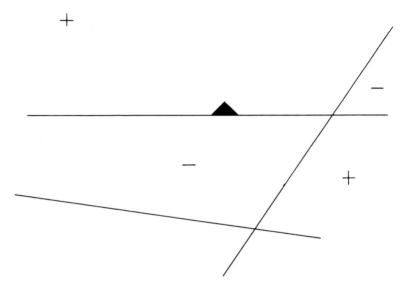

20. Schematic representation of large-scale + and − aggregates.

Figure 21 is to be thought of as a schematic representation of the universe on a scale much larger than the portion of the universe accessible to practical observation. Indeed, we are to think of our observations of all the galaxies, even the most remote ones visible in the largest telescopes, as occupying only a comparatively *small element* of just one of the aggregates of figure 20; for definiteness, let us say a positive aggregate.

The interactions on a particle anywhere in the universe will in this picture be a complicated addition of contributions from all the various aggregates. If we make the sensible assumption that, *on the average,* negative aggregates are as important as positive aggregates, the combined effect of all interactions at an arbitrary place is as likely to be negative as positive. Regions where the contributions add to a positive total will be separated from regions with a negative

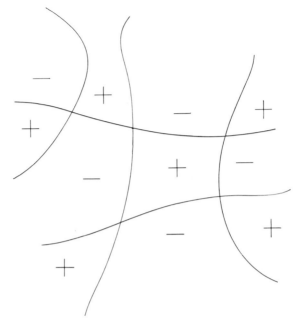

21. Our studies of cosmology may be concerned with only a small element of a much vaster universe.

total by surfaces on which the + and − contributions just cancel each other. The mass of a particle at the points of such surfaces will be zero.

In an extremely general way, we have now arrived at an understanding of the moment $\tau=0$. This moment defines a portion, perhaps only a small portion, of one of the surfaces of figure 20. The essential lesson we learn from the study of the red shifts of the galaxies is that *we happen, in our element of the universe, to lie near a surface of zero mass.* Notice that the surface in question need not be a single, unique surface, as we formerly took it to be. To emphasize this point, let us redraw figure 20 to show the

On the Origin of the Universe

range of astronomical observation as covering only a small element of a much vaster universe. This we can see in figure 21, where we end this lecture with the concept of a universe much greater in scale than any which has hitherto been considered in astronomy.

FOOTNOTES

[1] The reader concerned with technical details, and who is familiar with the usual cosmological theories, will recall that the frequency of monochromatic radiation varies as Q^{-1}, while the energy density of a homogeneous radiation field varies as Q^{-4}. In the usual theories, Q, the scale factor of the galaxy polygons, is variable. In particular, Q goes to zero at $t=0$, and so both frequencies and the energy density go to infinity. Thus, in the usual theories, the radiation field has very high temperature near $t=0$. In the present treatment, however, there is no change of Q, and hence radiation propagates without change of frequency (except as the radiation interacts with particles) and without change of energy density. Thus, the microwave background radiation, once generated near $\tau=0$, propagates essentially without change of frequency or energy. The concept that, near $\tau=0$, quanta of the radiation field are "hard" compared to particle energies, still holds good, however, because, near $\tau=0$, the particle masses are very small. Also, because of their small masses, particles near $\tau=0$ are effective "thermalizers" of the radiation produced before $\tau=0$ by stars on the "other side."

[2] Note that $(-)\times(-)$ is positive.

RADIOCARBON DATING

WILLARD F. LIBBY *For discovering and developing the radiocarbon dating technique, Dr. Libby received the 1960 Nobel Prize for Chemistry. Radiocarbon dating, a method of closely estimating the length of time since the death of ancient organic matter as much as forty thousand years old, was developed by Libby in 1947. The age of the organic sample is approximated by measuring its current rate of carbon-14 decay. Libby's dating technique has enabled man to explore the past as never before, establishing chronologies for recent geologic events, world climate changes, and man's prehistoric development.*

A physical scientist and educator, Libby has been Professor of Chemistry at the University of California at Los Angeles since 1959, and Director of the University of California's Institute of Geophysics and Planetary Physics since 1962. During World War II, Libby took part in the Manhattan Project, which developed the atomic bomb. His radiocarbon dating technique was developed and first applied at the University of Chicago and its Institute of Nuclear Studies (now the Enrico Fermi Institute for Nuclear Studies), where he taught after the Second World War.

Asked to serve on the U. S. Atomic Energy Commission by President Eisenhower in 1954, Libby served until 1959 and was a U.S. representative at the first and second International Conferences on the Peaceful Uses of Atomic Energy, held in Geneva, Switzerland, in 1955 and 1958.

I

RADIOCARBON DATING consists of the determination of the radiocarbon content of formerly living matter, and measures the time elapsed since death. It works over a period of some forty thousand years, with an error of measurement of about one century in the period zero to twenty thousand years and somewhat larger for the older dates. About one ounce of wood or charcoal or other carbon-containing matter is needed. Smaller amounts can be used, but it pays to have enough material for duplicate measurements in case of any difficulties.

The principal hazard is contamination by modern or very ancient carbon. However, experience has taught us how to do a "laundry job" satisfactorily, and in most cases in which we measure dates from samples of different sorts from a given site agreement is found. We prefer charcoal or wood or cloth or hair, because these materials are relatively easily cleaned and have proved over the years to contain only the original carbon. There seems to be no possibility of contaminating them with unremovable material if normal precautions are taken. Such materials as shellac, paraffin wax, and tar are very difficult to remove but normally don't occur in freshly excavated materials. Rather, one finds plant rootlets and ordinary dirt, which can be removed with tweezers and with chemical treatments such as acidic and basic washes. In the case of bone, we measure bone protein; somewhat larger samples are required, since the amount of protein in bone may well be reduced to 1 or 2 per cent in old bones, due to its destruction by soil bacteria. Such reduction appears not to discriminate between the carbon isotopes to any appreciable degree in the remaining material, so we are able to determine the concentration of radiocarbon, which after all is the measurement we seek to make.

The method depends on the bombardment of the atmosphere by cosmic rays coming from outer space and in particular from outside the solar system, although some production is accomplished also by solar cosmic rays. Cosmic rays undergo their first collision in about the top tenth of the atmosphere. These extremely energetic collisions shatter the atoms, producing free neutrons. The neutrons produced are captured by the nitrogen in the air, in particular by N^{14}, the abundant isotope, 99 per cent of the element, to make carbon 14 and an ordinary hydrogen atom. Also, in about 1 per cent of the cases, they produce C^{12}, which is stable, and the radioactive form of hydrogen called tritium. Tritium has a half life of twelve years, in contrast to radiocarbon with a half life of 5,730 years. Tritium dating has proved useful in measuring the age of wines and bodies of water. Even though it has a very different time scale, the principles are similar to those of radiocarbon dating.

Returning to the principles of radiocarbon dating, they are, in brief, that contact with the cosmic rays ceases at the time of death. Prior to that time and all during life, a direct connection is maintained through the food supply and the food cycle. The C^{14} produced high in the stratosphere by the energetic cosmic rays finds its way into plants and through plants into all animal life. The C^{14} atoms burn in the atmosphere probably almost immediately after production to form radioactive carbon dioxide, and this mixes with the ordinary carbon dioxide in the atmosphere, which is the food for all life. Plants grow on atmospheric carbon dioxide, moisture, sunlight, and salts. So, the assimilation of radioactive C^{14} in the food nicely balances the rate of decay of the stored C^{14} in the living organism. This entire system of animal and plant life, the CO_2 in the air, and most important of all the CO_2 dissolved as carbonate and bicarbonate in the ocean, is stirred rapidly in times short relative to the half life of radiocarbon. Thus, this system is a giant reservoir including all living beings as part of it.

At death, the connection is severed and the radiocarbon disappears at its own, immutable rate: 50 per cent every 5,730 years. The measurement of the content of radiocarbon per gram of carbon contained relative to that in living material gives the time since death: one half giving 5,730 years, one quarter giving 11,460 years, et cetera. This is the simple, basic principle of radiocarbon dating.

The atmosphere is stirred rapidly, as are the oceans, relative to the radiocarbon half life, so at any given time the radiocarbon concentration in living matter is uniform over the entire world. This was shown directly by E. C. Anderson in his doctoral thesis in the late 1940s at the University of Chicago and by several studies of atmospheric atomic explosions. The explosions release large quantities of neutrons, producing substantial amounts of radiocarbon in the atmosphere, which is measured as it mixes, the rate of mixing thus being determined directly. Consequently we have the firmest evidence that the radiocarbon concentration of living matter is uniform over the entire earth.

One basic assumption of radiocarbon dating is that the cosmic-ray intensity at the present time is the same as it was when the ancient organism died. This assumption has been shown in recent years to be a few per cent in error, and it is part of the modern radiocarbon dating technique to make a small correction for the change in concentration of radiocarbon in living matter as a function of time. Such a correction curve has been determined now for the past eight thousand years, and we use this curve. The correction is made by the use of a stand of trees at ten thousand feet altitude in the southern Sierra Nevada, the so-called Schulman Grove of bristlecone pine trees. This fabulous tree lives for four to five thousand years, and in this particular grove there are pieces lying on the ground that first fell there five thousand years or so ago. By comparing the outer rings of the fallen with the inner rings of the living, the entire ring chronology has been extended back to eight thousand years. The bristlecone has very thin rings and is sensitive to the variations of climate and living conditions, so that the pattern of ring thickness versus time is unique, and, like fingerprints, can identify a particular past span of time. For example, for two trees whose life spans overlapped in time, portions of the ring patterns are found to be identical. This wonderful dendrochronological work was performed by C. W. Ferguson and others at the University of Arizona Laboratory for Tree Ring Research. Thus it is possible to say to the nearest year how old a particular piece of wood is. In this way, one can obtain samples of wood of rigorously known age, and these samples, when compared with the radiocarbon date from the same wood, gives us the correction for the original concentration of living matter at the particular time. The correction curve has a broad general trend, as shown in figures 1 and 2, which indicate that the cosmic rays were more intense five to seven thousand years ago by about 5–10 per cent but it also shows that there are short-time fluctuations of fascinating interest. It seems very likely that the general over-all rise is due to a weakening of the earth's magnetic

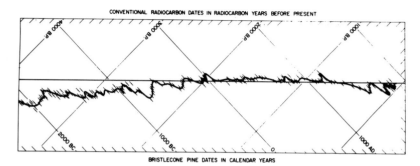

1. Bristlecone calibration curve.

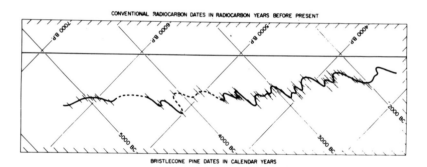

2. Bristlecone calibration curve.

field, which used to deflect fewer of the cosmic rays from striking the atmosphere. The wiggles must be due to something else. It is generally thought that they may be due to changes in the solar-wind intensity and thus might in the future tell us something about the history of the solar climate. Our colleague Dr. James Arnold, in separate research has shown by studying the cosmic-ray-induced radioactivities in meteorites, that the true cosmic-ray intensity outside the earth's magnetic field has remained approximately constant for many millions of years.

Radiocarbon Dating

To summarize, radiocarbon dating consists of the determination of the concentration of radiocarbon in dead organic matter formerly alive, the comparison of it with the concentration in living matter at the time of testing, and the computation from the simple logarithmic formula of the time spent in death. It is possible, in a radiocarbon dating lab, to measure three or four samples a week, with a duration time of about two weeks for each measurement.

Figure 3 shows the "Curve of Knowns," which we obtained early in the research. It indicates fair agreement, considering the rather large experimental error of measurement, but even at this early point one can notice that there was a tendency for the Egyptian samples to be too young.

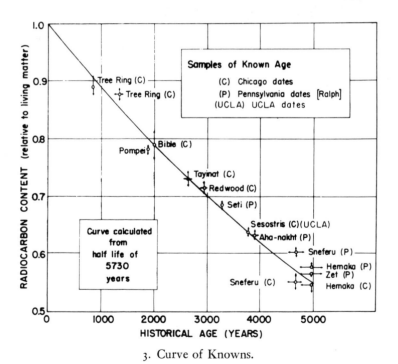

3. Curve of Knowns.

4. Linen wrapping from the Book of Isaiah Dead Sea Scroll, and strands of an Egyptian lady's hair.

5. Egyptian linen.

This was later noticed by Dr. Paul Damon at the University of Arizona, and subsequent research has resulted in the curve of corrections shown in figures 1 and 2.

Figures 4, 5, 6, and 7 are pictures of typical samples used. Figure 4 is of the linen wrapping of the Book of Isaiah from the Dead Sea Scrolls, and human hair five thousand years old from Egypt. Figure 5 is a linen handkerchief or cloth from a four-thousand-year-old tomb in Egypt. Figure 6 is a piece of rope from Peru twenty-five

6. Peruvian rope.

7. Oregon sandals.

hundred years old, and figure 7 is a sandal ninety-three hundred years old from a cave in eastern Oregon. These are typical of the types of material used. Perhaps most of the samples are less dramatic in appearance than these, but the principles are the same for all.

II

The method for radiocarbon dating was first disclosed in an article in *Science* in May 1947. In March of 1949 the first date list was published, also in *Science* magazine. The May 1947 article disclosed the discovery of natural radiocarbon in Baltimore sewage. It was followed by a more complete article in the November 1949 *Physical Review*. Actually, the research on the method began about two years earlier, in the fall of 1945, when I went to the University of Chicago as a young professor to what is now known as the Enrico Fermi Institute.

In the years before World War II, while I was a staff member at the Chemistry Department of the University of California at Berkeley and working on artificially produced radioisotopes, I had a graduate student searching for radiocarbon. It had been shown earlier by Dr. Harkins of the University of Chicago and his students, in particular Franz Kurie, Martin Kamen, and others, that nitrogen when irradiated with thermal neutrons emitted protons. This seemed to indicate very definitely that a carbon isotope of mass 14 might exist. Sam Ruben, my graduate student, and I put some hundreds of pounds of ammonium nitrate near the cyclotron target in the Radiation Laboratory at Berkeley in such a way as to be irradiated by the neutrons coming from the target. We calculated from the measured absorption cross section of nitrogen for thermal neutrons, making the erroneous assumption that the half life of the unknown isotope would be about three months (in comparison to what we know now to be 5,730 years) and that a detectable amount of radiocarbon in the ammonium nitrate would be produced in some six months of irradiation—but we found nothing. The technique we used was later proved to be impeccable. The only problem was that we had underestimated the half life, and therefore far too little radiocarbon was produced to have a detectable radioactivity. Later, in 1942, Ruben and Kamen tried again, this time more forcefully, bombarding graphite with deuterons, the thought being that deuterons hitting the stable isotope natural C^{13} would make C^{14} plus a hydrogen atom in the graphite. This proved to happen, and so radiocarbon was discovered.

In 1939, Professor Serge Korff of New York University had published his finding of neutron production in the high layers of the atmosphere, which he had discovered by sending a neutron-sensitive counter up in a balloon. This remarkable and epic-making piece of research, together with the later work of Ruben and Kamen, clearly indicated that nature should contain radiocarbon. Busy on World War II problems, we did nothing much more until the fall of 1945, when we set busily to work to determine two things, the first being the half life of C^{14}. Values as high as the Ruben-Kamen life of twenty-five thousand years were being quoted. Obviously, for radiocarbon dating purposes it was essential that a good value be available. Fortunately, others were working on the problem at the same time. It has never been clear to me why they worked on it except that it is true that radiocarbon is quite extraordinary in its long lifetime, considering the energy of its beta-ray emission compared with those of other radionuclear species. Ruben and I had used sulfur 35 as our model. It has an eighty-seven-day half life and has almost exactly the same energy radiation as does C^{14}. So, apparently this nuclear-physics enigma C^{14} had intrigued others to measure its half life more accurately. We worked at the Argonne National Laboratory in Chicago for something over a year, making careful measurements of the half life, obtaining synthetic C^{14} from the Atomic Energy Commission. Dr. Mark Inghram of the physics department in Chicago measured its concentration in these specimens for us with his mass spectrometer. Diluting a carefully measured concentrated $C^{14}O_2$ to a low enough level, making absolute-count measurements, correcting for wall losses and end losses by using counters of various lengths and diameters, we determined that the half life was 5,568 years. The first radiocarbon dates were published with 5,568 years as the base half life. This is still the practice today (although since then the half life has been more accurately determined) because of the confusion that would result if dates were computed using a different half life. Thus, a radiocarbon date published today must be corrected to the new half life by multiplying by 1.03, for a 3 per cent increase in age.

The second aim we had was to move quietly and carefully and cautiously to test the dating idea. The first thing to test was whether or not living matter contained radiocarbon. This was accomplished working with Dr. A. V. Grosse, then with the Houdry Process Corporation, whom I had met during the Manhattan District work. Dr. Grosse had an apparatus for producing C^{13}-enriched materials to be used in cancer research. We got samples of living methane (methane from a biological material), namely Baltimore city sewage, and enriched it in Dr. Grosse's thermal diffusion column. We then measured the enrichment of C^{13} in the sewage methane and looked for a detectable additional count rate of C^{14} in counters filled with this methane as compared with that in unenriched methane obtained from oil wells, where the C^{14} would have long since decayed due to its age. We found a positive excess indicating an abundance close to that predicted on the basis of Dr. Korff's cosmic-ray balloon measurements together with our computation of the total amount of carbon in equilibrium with atmospheric CO_2.

The latter computation was made assuming that the oceans are mixed in a time short compared to the life of radiocarbon. We came to this conclusion in a number of theoretical ways, the main argument being that, if nothing else mixed the oceans, the thermal heat emitted from the ocean bottom would do it. Later, it was possible to radiocarbon-date deep ocean waters and show that the mixing time of the ocean is about fifteen hundred years. Taking into consideration the ocean carbon (which is mainly the sea salts carbonate and bicarbonate), the organic matter in the ocean, the living matter in the ocean and on the land,

and the atmospheric CO_2, we estimated a total of about eight grams of carbon per square centimeter of earth's surface, which is mixed with the atmospheric CO_2. Thus Dr. Korff's neutron data allowed us to predict the C^{14} concentration as being equivalent to two disintegrations per second per eight grams of living carbon. All this assumes the cosmic rays to have been constant in intensity over the past several tens of thousands of years.

It seemed reasonable also that the mixing would be so rapid that even though the cosmic-ray intensity was strongly latitude-dependent, the mean concentration of radiocarbon in living matter world wide should have the same C^{14} concentration. Dr. E. C. Anderson tested this; his results, which were published in the March 1949 *Science* article together with the first date list showed world-wide constancy. Thus, our basic mixing postulates appeared to be correct. It also seemed that the assumption of the constancy of the cosmic rays must be nearly right, since we found a concentration in living matter that was close to that which we estimated from present-day cosmic-ray neutron intensities. This allowed us to assume that the concentration in past times for living matter had been the same as it is today. As I've said earlier, this is only about 97 per cent true, and part of the history of radiocarbon dating involves the discovery of methods of determining the degree to which deviation occurs. At this time, however, we were ready to proceed with the dating, but we had a very difficult problem to solve.

Namely, Dr. Grosse's apparatus was elaborate; so, the enrichment of a single sample was so expensive that it would have been difficult to justify the method to archaeologists, who are notoriously underfinanced in most cases. So, instead, Dr. Anderson and I spent some months trying to devise a technique that would reduce the background count rate in our counters and thereby allow unenriched samples to be measured. Our original method of measuring the samples Dr. Grosse had given us was simply to put the unenriched methane gas in a Geiger counter and observe the rate as compared to the count rate of a similar counter filled with dead methane from petroleum wells. This method is far too insensitive to measure the radiocarbon concentration in unenriched methane to reasonable accuracy, so that we would have had to depend on enriching every sample. At first we installed a four-story-tall isotope-enriching apparatus in Kent Hall on the University of Chicago campus, using funds from the Axel Wenner-Gren Foundation (then known as the Viking Fund), which supported our work on radiocarbon dating together with the U. S. Air Force Office of Scientific Research, but, after all, this apparatus was used only for practice.

During all of this work we did not announce our objective. We merely said we were determining the half life of radiocarbon and working on natural cosmic-ray isotopes, because we were afraid that such an idea would not be well received and supported, and it was only after our first radiocarbon dates were in hand that the method was announced. Our caution may have been a carry-over from the habits of the Second World War, when secrecy on the Manhattan Project was the word, rule, and law, for reasons of national defense. However, then as now, a relatively unknown young scientist has to be very careful in his solicitation of support for his research. He must not appear to be too wild and presumptively hopeful. Otherwise the referees will turn him down. I had said I wanted to work on the problem of the half life and the natural occurrence, but I believed and still do that had I said my plan was to trace human history through cosmic rays I would have failed to obtain support.

In any case, we had to choose between using Dr. Grosse's column, which would make our method far too expensive to be of value to archaeologists, and measuring in a deep mine, where the cosmic rays would be eliminated. There were no mines near Chicago, the water table being only ten feet below the surface of the ground, so we tried to invent something to get us out of our difficulty. Over the years, we had followed the development of counting technology. We built the first Geiger counter in the United States, in my laboratory at Berkeley in the 1930s. In particular, we learned about the technique of connecting counters in electrical coincidence in order to determine the direction of travel of cosmic rays. We were piling large amounts of shielding around our Geiger counter and finding that six inches of iron or lead would eliminate all counts due to uranium, thorium, potassium, and ordinary radioactivity in the walls and floors of the building, but the residual cosmic-ray background of about one count per minute per square centimeter cross section of the counter was essentially uneliminable by even many feet of steel. We actually moved our counter and placed it underneath the huge steel yoke of the cyclotron then being built at the University of Chicago, hoping that in this way the ten feet of steel above it would appreciably reduce the cosmic-ray count rate. It reduced it perhaps 30 per cent. We were fairly discouraged about our prospects until finally we solved the problem. The solution was to place around the counter with the archaeological sample in it a set of Geiger counters that touched each other, so that any penetrating cosmic ray that went through the carbon-dating counter would certainly have passed through one of the shielding counters. We then arranged that the shielding counters turn off the carbon-dating counter for about one thousandth of a second when a cosmic ray passed through, thus eliminating the cosmic rays from the record. The rates in our present counter at UCLA are about four thousand counts per minute without any shield, one thousand counts per minute with the iron shield, and about twelve counts per minute with the anti-coincidence shield connected (figure 8). In this way, it was possible to obtain clean, accurate count rates from the natural radiocarbon itself, without the added cosmic-ray count rate, and we were now able to measure unenriched radiocarbon in archaeological materials as old as thirty to forty thousand years.

At the present, in our counters, we use CO_2 gas obtained by burning the sample, but at that time it was not known that CO_2 gas could be used in counters. It in fact had a very bad reputation until Dr. Gordon Fergusson in New Zealand discovered that pure CO_2 was a good counter gas. I was told he had been trying fruitlessly for quite some time to purify CO_2 so it could be used in the counter when in desperation he took a fire extinguisher off the wall and used its CO_2 and found that it was indeed a good counter gas.

Returning to our experiences in 1947, we reduced our CO_2 with magnesium metal to elementary carbon, extracted the magnesium oxide from the carbon black with acid, wetted it with ethyl alcohol, and plastered the carbon black on the inside of a brass cylinder. We spread it out over about three hundred square centimeters of area on half of a cylinder twelve inches long and about three inches in diameter. This then served as the wall of a Geiger counter, the entire cylinder being surrounded by a vacuum-tight chamber down the center of which a central wire passed. We made the cylinder twice as long as the active part of the wire so that, by moving the cylinder, we could measure background and thus, by subtraction, meas-

8. Anti-coincidence-shield counter.

ure the count due to the carbon sample (the chamber was one and a half times the length of the cylinder). Figure 8 shows this counter. We obtained with modern wood a difference of merely four counts a minute above a background count of about ten. That is, with the uncoated brass cylinder over the sensitive portion of the wire we would obtain ten counts a minute and with modern carbon coat fourteen counts per minute. By counting overnight and watching and controlling very carefully the behavior of the electronics, we were able to achieve an accuracy adequate to measure some dates. In this way, the Anderson assays and the first carbon dates were measured and published. In fact, we used this method to obtain some seven hundred radiocarbon dates. These carbon blacks were stored in bottles after the measurements were made, and many of them are still filed in our laboratory at the University of California, Los Angeles. It has been proposed that some of the measurements be remade with modern techniques, for the accuracy of the early carbon dates left something to be desired.

The next step was to try the dating method. Dr. James Arnold of Princeton joined us for these tests. He was a physical chemist, like Dr. Anderson and myself, but his father, a lawyer, was an enthusiastic amateur archaeologist, and this brought him to us in an enthusiastic mood. We immediately asked, "How can we coax a museum keeper to give precious and valuable materials for us to burn up?" He worried about this with us, and finally we decided that we had to enlist the aid of recognized experts to advise and help acquire the materials. So we appealed to the American Archaeological Association and the Geological Society of America to appoint for this purpose a committee of experts, which they did. The chairman was Frederick Johnson of the Peabody Foundation in Andover, and other members were Professor Froelich Rainey of the University of Pennsylvania Museum, Donald Collier of the Field Museum in Chicago, and Richard Foster Flint, the geologist from Yale. These gentlemen arranged for us to work with Dr. John Wilson, a senior professor at the Oriental Institute at the University of Chicago, and through Professor Wilson we obtained materials from the earliest Egyptian pyramids and proceeded to burn and date them and compare the results with the ages known by Egyptologists. The agreements were well within the counting uncertainty of a few centuries, and we were highly elated.

However, we had a frustrating experience in that the third sample given us by Dr. Wilson turned out to be modern! We were pretty discouraged by this and nearly gave up the method, until Dr. Wilson, I believe it was, suggested that maybe our answer was right and that Dr. Breasted, the great collector of the Oriental Institute Egyptian material, either had made some repairs on the particular coffin using modern wood or else it was somehow a fake. In any case, we went forward, and now, some twenty-five years later, during which many tens of thousands of dates have been measured in dozens of laboratories, we can take a reviewing look at the method. By the way, our original counting equipment is in the collections of the Smithsonian Institution.

III

I will speak of the results in a personal way as a physical chemist studying archaeology under the careful tutelage of very great experts. More appropriately, I will tell of the contributions to our understanding of human history and the history of the earth and geophysics in general.

The dates are published semiannually in the magazine *Radiocarbon*, edited by Edward S. Deevey, Richard Foster Flint, J. Gordon Ogden III, Irving Rouge, and Minze Stuiver; Managing Editor, Renée S. Kra, at Yale University. They have now compiled tens of thousands of dates from dozens of radiocarbon dating laboratories in some twenty countries. There have been eight international radiocarbon conferences, the last one in New Zealand two years ago with over one hundred attendees and with twenty-one countries represented. It is truly an international effort, with no central funding or direction, yet it is collaborative in the extreme. For instance, the convention to use the Chicago half life despite the fact that we all know that it is 3 per cent low, and the use of the National Bureau of

Standards oxalic acid to define the modern radiocarbon concentration, are conventions that have been arrived at by informal agreement on an international scale. This shows that international collaboration in scientific projects is possible when no seemingly direct practical application is involved. What a commentary that collaboration on more useful projects seems to be too difficult! In any case, there is an international radiocarbon community, which meets about every two or three years and discusses the recent results from radiocarbon dating, improvements in the method, and related subjects, such as tree-ring research.

The first results from the Chicago laboratory were the Egyptian dates, and the results were highly satisfying although, as remarked earlier, they seemed to be somewhat younger in the earliest dynasties than the historical dates. However, the uncertainties in the historical ages of the early dynasties were such that it was difficult at that time to be certain that there was any disagreement at all. Now that we have the bristlecone-pine chronology correction curve and can believe that the calibration that applies to the Schulman Grove in the southern Sierra Nevada applies world wide because of the rapid atmospheric mixing, we go back to look again. I refer particularly to the article by Dr. I. E. S. Edwards, the Keeper of Egyptian Antiquities in the British Museum, given at the symposium on the Impact of Natural Sciences on Archaeology held in London in December 1970, published in Volume 269 of the *Proceedings of the Royal Society of London*. He concludes that, using the tree-ring correction curve, 71 per cent of twenty-nine Egyptian dates in the past three-thousand-year period show difference from historic dates of only fifty years or less, and eighty-nine per cent are in agreement to within one hundred years or less. With this confirmation of our method, we gained confidence to look at the earlier dates for the first dynasties. We found that we confirmed the estimates of the historians and that their detective work was remarkably correct. Going back in time to predynastic Egypt, before the historical record, we measured at Chicago as recorded in our book *Radiocarbon Dating*[1] material simply labeled "Middle-Pre-Dynastic," sent to us by Ferdinand de Bono of the Cairo Museum. It was charcoal from a house floor at El Omari, near Cairo. Our date was six thousand years. Another predynastic Egyptian sample was obtained from grain from a granary submitted by Miss Caton-Thompson and Mrs. Elise Baumgartel of the University of Cambridge and described as from the Desert Fayum. This we dated at seven thousand years, or some two thousand years earlier than the Ist dynasty. I well recall eating some of this wheat. It tasted as though it were fresh. The great thrills Dr. Arnold and I had in laying our hands on some of these priceless materials have left memories that we shall never forget. I remember the dry dregs of beer in a beer stein found in one of the Ist-dynasty tombs.

Although we did not personally visit the pyramids, we came to feel a familiarity with them. Some years ago, Dr. Edwards and I arranged to collect archaeological samples just for radiocarbon dating. Prior to that, we had always depended on our friends' samples in the museums. We arranged to send a professional expedition to collect samples, and it was the dates for these samples that were discussed in Dr. Edwards's paper. They were carefully dated both at UCLA and at the British Museum radiocarbon laboratories.

It was of course inevitable that we would become emo-

[1] Willard F. Libby, *Radiocarbon Dating* (Chicago: University of Chicago Press, 1955).

tionally and personally involved in the archaeological excavations, for there is a basic human interest in the history of man, but Dr. Arnold and I never erred by questioning our archaeological and historical colleagues, except in the case of the Rameses sarcophagus with its modern radiocarbon date, which forced us to question its authenticity. Later, we dated the Dead Sea Scrolls, in particular the Book of Isaiah, using the linen wrappings. The results were in agreement with the assumption of authenticity. We actually put it on the Curve of Knowns for the reason that there was no doubt about the historic age deduced from the writing in the scroll. We dated other materials from the time of Christ, and in each instance obtained an excellent check. One of the reasons that led us to be somewhat slow to recognize the curve of deviations was that it just so happened that there are essentially no deviations in this period.

To continue with the archaeological story, we obtained samples from many parts of the world, particularly North America and Europe, and one of the most interesting and important of these is the Lascaux Cave in the Dordogne in central France. This material was furnished by Professor H. L. Movius of Harvard. Figures 9, 10, and 11 show some of the beautiful paintings found on the walls of this cave. Our sample was charcoal taken from the occupational level in the northwestern portion of the cave by Abbé H. Breuil and Monsieur Séverin Blanc. I well remember it being contained in an inconspicuous little matchbox wrapped in aluminum foil. This material dates at sixteen thousand years without any correction for the deviations due to cosmic-ray variations, since our tree-ring correction curve stops at eight thousand years and we therefore do not know the correction for sixteen thousand years.

It is our practice to quote dates in this way and call them "radiocarbon ages" in the hope that if we ever are able to

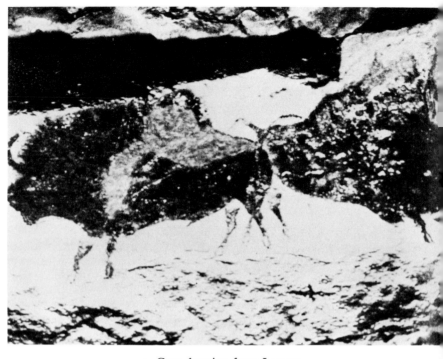

9. Cave drawing from Lascaux.

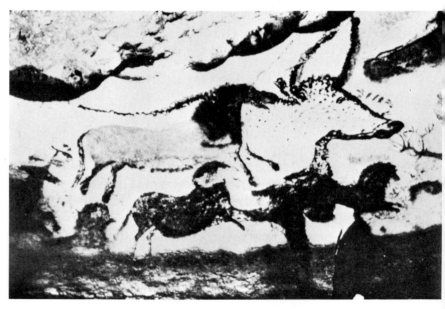

10. Cave drawing from Lascaux.

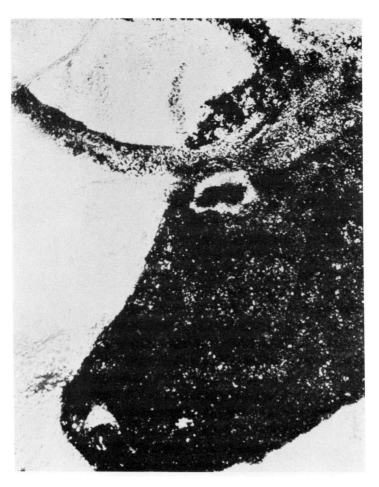

11. Cave drawing from Lascaux.

calibrate back beyond eight thousand years, we will be able to apply proper corrections to obtain absolute ages. So we say that approximately sixteen thousand years ago the cave was occupied by humans who made the paintings. However, Professor Movius believes that the paintings are apt to be older than the charcoal, I understand. We have but to compare these paintings with modern art to realize that as far as painting skills go, man has not changed substantially! The subtlety, grace, and beauty we associate with great art are obviously contained in these very ancient works. What a thrill to understand that! I suppose it was obvious to Professor Movius that they were of great antiquity, for he has published many articles implying it, but I know that the radiocarbon dates gave him comfort, reassured him, and gave him confidence to carry forward great studies on Cro-Magnon man and his appearance as Neanderthal fades out. My general impression is that Cro-Magnon appeared about thirty thousand years ago. As we can now date back to forty thousand years or so, depending on the availability of dating materials, it is possible in principle to establish this great event.

Actually, there is a kind of natural limitation to the age range in that with samples of such antiquity the problems of laundry are by no means negligible. At ten half lives, that is 57,300 years, the residual radiocarbon is 2^{10} lower, that is one part in 1,024, so if one has an impurity of 0.1 per cent of modern carbon, the radiocarbon age is fallacious. It is very difficult to be certain of purification to that degree. With ample supplies of materials and a great deal of careful washing, many samples can be measured nevertheless, so that dates that seem reliable back to forty thousand years or more have been obtained.

Continuing with archaeology, we turn our attention to England, to many samples of the greatest interest. I mention here the Starr Carr site in Yorkshire, a mesolithic site at Lake Pickering. The material was from a wooden platform submitted by Professor Harry Godwin of Cambridge University dated by us at ninety-eight hundred years. Although the curve of tree-ring correction ends at eight thousand years, we can guess that it is probably necessary to correct by adding another five hundred years. This date is postglacial and refers to the advent of first man in England following the glacial ice sheet, which had swept evidence of earlier, preglacial man away.

We see the discovery that man retreated before the advancing ice here and throughout England and northern Europe and Scandinavia, and then repopulated the glaciated regions as the ice melted. The oldest human artifacts are about this age. We were much worried about Piltdown Man, which, if authentic, obviously was preglacial, for his human bones were supposedly intermixed with elephant bones. When Kenneth Oakley showed it was a fake, we felt better. For reasons that we will discuss in a moment, it appears to be true also throughout the Americas that man followed the retreating ice sheet.

In North America, turning to the dating of the last ice age, we measured the Two Creeks forest-bed samples given us by Professor L. R. Wilson of the University of Massachusetts and Professor Harlen Bretz of the University of Chicago. The Two Creeks, Wisconsin, site is a forest bed that underlies the Valders glacial drift. Apparently, the spruce forest was submerged, pushed over, and buried by the last advancing ice sheet in this region. It is the definitive site for dating what is known as the Mankato, which is a term for the last glacial advance in North America. The pieces of wood and the peat in the soil in which the wood was found were dated in agreement at 11,800 years without any correction being applied, so probably twelve thousand years is a good round figure for the time of furthermost advance of the last ice age. We then turned our attention to Europe and measured similar sites and found similar dates. Recent measurements in New Zealand, though few in number, have indicated that glaciation occurred simultaneously there also. We assume therefore that it was a world-wide phenomenon and this pile-up of continental ice lowered the sea level by some 150 feet world wide. Now we find that the oldest men in North, Central, and South America date at about eleven thousand years, and the coincidence of their arrival with the retreat of the ice sheet is very strongly evident. It seems clear that at that time there was a great human invasion from Siberia along beach pathways now about 150 feet underwater, which came inland from the beaches at about Vancouver or Seattle.

The postglacial climate was very salubrious, in Nevada the rainfall amounting to some thirty inches per year as judged by the types of plant seeds found in the manure of the ground sloth, which was living in Nevada in large numbers at that time. He was an animal about the size of a horse, a grass eater. So what we now know as the great desert of the Southwest was a pleasant place ten to eleven thousand years ago. Then the weather changed, and many of our best samples have come from this region, for, like Egypt, they are beautifully preserved due to the hot, dry atmosphere that has obtained in the past eight thousand years or so.

I recall a corpse found in a Nevada cave that was twenty-five hundred years old and essentially perfect. The lady had not been mummified. She was merely laid on a rug in a dry cave, and except for a portion of her body that had been chewed away by rats, her skin, hair, and clothing were all essentially intact. This specimen is in the Santa Barbara Museum of Natural History.

Returning to the earliest man's story, we measured dates of great antiquity using archaeological samples furnished us by Professor L. S. Cressman of the University of Oregon. Professor Cressman sent us several pairs of woven rope sandals from a cache of some three hundred pairs found in a cave called Fort Rock by a road crew. (One of these is shown in figure 7.) Fort Rock Cave was buried beneath pumice from the eruption of the volcano Mount New-berry. The sandals date at ninety-eight hundred years, sim-ilar in age to the Starr Carr site in England. There are many other old human living sites, throughout North, Central, and South America, showing that the early Amer-ican Indian was roaming everywhere at ten thousand years. The Danger cave site is in western Utah, at the Nevada-Utah line, at Wendover, which is 120 miles west of Great Salt Lake. It is one hundred feet above lake level at present but has on its floor, beach sand two feet thick in which driftwood is found together with sheep droppings, on top of which human debris some ten or more feet thick is found. This constitutes a historical record of man from some eleven thousand years ago to the present. The cave debris was furnished us by Professor Jesse Jennings of the University of Utah. Professor Cressman's monumental work *Archae-ological Researches in the Northern Great Basin*[2] postu-lated the antiquity of the sandal wearers in Oregon, and our dates confirmed his estimates.

Mount Newberry erupted again some two thousand years ago, after lying dormant for about eight thousand years. The charcoal produced by the hot pumice falling on forests is ideal material for radiocarbon dating, making it relatively straightforward to determine the time interval between eruptions of volcanoes. We are trying to evalu-ate the history of great earthquakes on the San Andreas Fault in California, but it is less obvious how to identify the landslides and dams made by earth movements. We know there are such movements from modern experience, but so far it has been very difficult to get useful samples for carbon dating of earthquakes as accurate as those for volcanoes.

We collaborated on sites in Alaska with Professor Froe-

---

[2] Jesse Jennings, *Archaeological Researches in the Northern Great Basin*, Publication 538 (Washington, D.C.: Carnegie Institute).

lich Rainey, finding that the human remains there were relatively young. However, I understand that recently obtained dates have shown that man may have been present before the last glaciation.

Our general finding that humans have been in continuous occupation of the Americas for the past eleven thousand years does not exclude the possibility that they were here earlier. It is just that we have few if any radiocarbon dates indicating preglacial human activity. I well recall a finding that some charcoal from a site at Tule Springs, Nevada, near Las Vegas dated at twenty-five thousand years, suggested the presence of man at that time. This site had been excavated by Fenley Hunter and M. R. Harrington of the Southwest Museum in Los Angeles, both very reputable archaeologists. The sample consisted of charcoal taken from beneath an ash bed about two feet below the present surface. This exciting discovery led us to excavate some twenty-five square miles around this site because of the great importance preglacial man has in archaeology. When I went to the University of California, Los Angeles, after leaving the Atomic Energy Commission, in 1959, in addition to setting up our Radiocarbon Dating Laboratory I organized a little non-profit foundation to assist in special archaeological excavations and other good works. One of the members of our Isotope Foundation, the president of a construction firm, Herschel Smith, offered to lay on bulldozers and labor free, for digging the great trenches the archaeologists wanted. To make a long story short, for some fifty thousand dollars that the National Science Foundation furnished, together with a little money from our Isotope Foundation, we performed what we estimated to be about four hundred thousand dollars' worth of work with dozens of graduate students and the enthusiastic and essentially full-time services of Dr. Richard Shutler, then of the University of Nevada and now of the University of Iowa. The findings fill some eight volumes and constitute one of the largest efforts ever made in archaeology, but what we found was that there was no man at this site earlier than eleven thousand years ago, although there were many kinds of animals going back as far in time as we could date with radiocarbon. We obtained in this way, however, a fascinating record of the history of the climate. Finally, it turned out that the "charcoal" was not charcoal. I am ashamed to admit this, for I should have known better, being a chemist. There is a simple chemical test, which I should have used and did later. Apparently the "charcoal" was wood on its way to becoming coal, peat, and lignite, and it was indeed twenty-five thousand years old, but it wasn't charcoal (figure 12). Normally, one is certain that charcoal is an indicator of man, providing it is found in the proper context in a circular deposit indicating a campfire or in a cave so that it could not have resulted from a forest fire, so the charcoal finding of Dr. Harrington led to this exciting but abortive research, which taught us to be more cautious.

In another instance, the late Dr. Louis Leakey, the famous early-man archaeologist from Kenya, when visiting California, told me that there was a skull in the Museum of Man in Paris that supposedly came from California and that he wished us to date it. I should mention here that one of the important developments in the past twenty years of radiocarbon dating technique was the perfection of the method of bone dating using the protein from the bone, which although small in amount percentagewise often is the only material useful to date a site. This skull dated at something around eighteen thousand years, and so once again we organized a dig. The site was Laguna Beach, in Southern California. We obtained permission of the city fathers, rented the particular property where the skull had

12. Preglacial wood.

supposedly been found, and put up a fence. Professor James R. Sackett of the Anthropology Department at UCLA and my colleague Professor Rainer Berger, who operates the UCLA Radiocarbon Dating Laboratory, together with students, searched for months for evidence of human burials, with entirely negative results.

Dr. Leakey died thinking that he had strong evidence for preglacial man in a desert site known as Calico, in the southwestern desert. This site contained stones that he thought were tools almost certainly shaped by human hands. Other archaeologists were less convinced, so I would say that, at the moment, the whole question of preglacial man in the Americans remains open. If any appropriate material is ever obtained, we will be able to date it, providing that it contains carbon. The search continues and the archaeological community is divided as to what to expect. It seems clear that the lowering of the sea level during the last ice age exposed beaches, which are now 150 feet under water, on which it was possible to walk from Siberia to the vicinity of Vancouver or Seattle. There is no doubt that there is a great deal of fascinating archaeological material presently submerged on former beaches the world over. In La Jolla, California, for example, we know there are thousands of pots about one hundred feet down below sea level, and a ten-thousand-year-old man was found buried in the cliffs there. So there is no doubt that man was ubiquitous ten thousand years ago in the Americas. However, there were three preceding ice ages, and it is strange that Asian man didn't walk across then. Perhaps he did but was too few in number to leave enough remains to survive the years with reasonable probability. We've recently tried to mount a collaborative research with the Chinese to investigate very ancient man in Mongolia, to obtain evidence of eleven-to-twelve-thousand-year-old man there.

Cro-Magnon man, the species to which we ourselves belong, dates back to something like twenty-five to thirty thousand years in Europe and may be somewhat older. He appears to emerge suddenly as a new entity. There remains much detail to be fitted into his great story. There is hope that radiocarbon dating will be able to do that, if we can find datable samples.

We are now organizing an international program for evaluation of the history of recent climate. It is possible, by dating trees and other materials, to establish the history of the climate back to the time limits of radiocarbon with considerable accuracy. Other dating methods can be used for earlier times but unfortunately tend to be somewhat less accurate, at least at their present stage of development. This program, known as the "International Isotope Decade," it is hoped will concentrate on the world-wide manifestations of the onset of the last ice age, its transition into the postglacial period, and the ups and downs of the climate so dramatically manifested in the southwestern United States and in North Africa in the great deserts and dry lakes. We believe that if we evaluate this information we may be able to tell what caused the glaciation and from this to predict future ice ages.

As you all know, our food supply is closely linked to the climate and we are in a precarious situation as far as food is concerned. We desperately need to predict the weather changes over periods of the next years. This is our hope. Radiocarbon has usefulness in geophysics and the history of geophysical processes, and we are most anxious to develop this aspect at least as fully as we have done in the more obvious applications to archaeology.

THE MOON AS
THRESHOLD

ISAAC ASIMOV *Russian-born American biochemist and science-fiction writer Isaac Asimov published his first story in 1938, while attending Columbia University. Since that time, he has written more books on a wider range of themes than any author in history, as well as numerous short stories, essays, and occasional television scripts. Best known for* I, ROBOT *(1950), the* FOUNDATION *trilogy (*FOUNDATION, *1951;* FOUNDATION AND EMPIRE, *1952; and* SECOND FOUNDATION, *1953) and other science-fiction works, Asimov is also a leading writer of popular science texts. His 163 books range in theme from ancient history, to a guide to the Bible, to a two-volume guide to Shakespeare, and annotations of* DON JUAN *and* PARADISE LOST.

Asimov is an Associate Professor of Biochemistry at the Boston University School of Medicine, where he taught full-time until 1958. He is coauthor of the medical textbook BIOCHEMISTRY AND HUMAN METABOLISM *(1957).*

THE TWENTIETH CENTURY has seen mankind complete its conquest of the Earth.

On April 6, 1909, the American explorer Robert Edwin Peary reached the North Pole. That meant human beings could range as far north on the face of the globe as they thought it advisable to.

On December 16, 1911, the Norwegian explorer Roald E. G. Amundsen reached the South Pole and extended man's range to the southernmost extreme.

On May 29, 1953, the New Zealand explorer Edmund Percival Hillary and his Sherpa colleague Tenzing Norgay set foot upon the top of Mount Everest, that point on Earth farthest above sea level, and mankind's range was extended to the maximum altitude of the solid surface of the planet.

On January 23, 1960, the Frenchman Jacques Piccard and the American Don Walsh descended by bathyscaphe to the bottom of the Marianas Trench, in the western Pacific Ocean, and mankind's range was extended to the deepest abyss to which Earth's solid surface plunges.

If there remained any point on Earth's surface, north or south, high or low, that had not felt a human footstep, it was only because no one had yet taken sufficient trouble to reach that point.

But the most remarkable extension of the human range witnessed by the twentieth century has been upward, away from the surface of the Earth. The man-carrying balloon was invented in 1783, and till the very end of the 1800s remained the sole medium by which men could explore the open atmosphere.

On July 2, 1900, however, a human being first rose from the surface of the Earth in *powered* flight when the German inventor Ferdinand von Zeppelin, sent up the first dirigible, and on December 17, 1903, the American inventors Wilbur and Orville Wright flew the first heavier-than-air flying machine, the airplane. Within half a century, planes were penetrating the stratosphere and moving at speeds greater than that of sound. The lowermost twenty-odd miles of the atmosphere was open to man—and that was not the limit, either.

On October 4, 1957, the Soviet satellite Sputnik I went into orbit about the Earth at heights where only traces of the atmosphere existed, and mankind had made its first move to penetrate space. The first *manned* satellite, carrying the Soviet cosmonaut Yuri Alekseyevich Gagarin was placed in orbit about the Earth on April 12, 1961, and on July 20, 1969, the American astronaut Neil Armstrong became the first man to set foot upon a world other than the Earth—on our natural satellite, the Moon.

It has been a remarkable leap: from the first powered flight to the landing on the Moon in sixty-nine years and eighteen days. That represents a lapse of years less than that of a single normal life—that of my father, for instance, who was born three and a half years before the flight of the first dirigible and died sixteen days after the first Moon landing.

The question is, where can human beings go next? Can the record of the steady outward extension of humanity's range continue? Can the Moon, having been reached, serve as a threshold to farther regions of space and to conquests far more staggering than anything that has preceded?

Without direct human participation, the exploration of space has extended beyond the Moon, and even much beyond it. Unmanned probes have been sent out into space and have sent back information from successively more distant bodies—from Venus, Mars, Mercury, and Jupiter. As I write, Pioneer 11 is racing toward a rendezvous with Saturn and will, we hope, send back photographs at close range of this farthest of all planets known to the ancients.

The far penetration of human instruments, without human beings themselves involved, does not, however, have the glorious ring of accomplishment that we associate with the mystique of exploration. Plumb lines may be dropped to the ocean depths, but that does not have the glory of men in bathyscaphes doing the same. Drills have penetrated more deeply into the solid Earth than human beings can go in the foreseeable future, but that scarcely counts as an extension of range.

Never mind mankind's probes; where can *mankind* go beyond the Moon?

It may be that we will be forced to give a disappointing answer here, for though the lunar landing was a great triumph, the achievement was a deceptively simple one. Earth and Moon are a double world, which occupy what can only be considered neighboring points on any but the very smallest astronomic scale. They are separated by only 380,000 kilometers, a distance equal to only ten times the circumference of the Earth and one that anyone driving a car under average American conditions can exceed in a lifetime of stopping and starting.

Look at it this way. It takes only three days to reach the Moon; and it took seven weeks for Columbus to reach the New World. In reaching the Moon, we have made only the most microscopic dent in the vastness of space. Indeed, we have not really left Earth, since the Moon is as much a slave to Earth's gravitational influence as an apple on a tree is—something Newton perceived three centuries ago.

In going beyond the Moon, we must take our first really large step, and here the easy victories are over at once.

The nearest sizable body other than the Moon is the planet Venus. Even when it is at its closest to Earth, it is 40 million kilometers away, and that is 105 times the distance of the Moon.

We don't expect a space vessel to move straight across the gap between the planetary orbits. The most economical route for the space vessel to follow is an elliptical orbit of its own that begins at Earth and intersects the orbit of Venus just as that planet is approaching the intersection point (which makes for an intricately calculated trajectory). The space vessel must travel a total distance far greater than the minimum separation of the two planets, and the voyage will take some six months.

Already, men have remained in space as long as three months, but that was in Skylab, in Earth's immediate neighborhood, where rescue at short notice was possible. To spend twice the time, with every moment taking you farther from home, is a psychological hazard indeed.

Worse yet, having arrived in the neighborhood of Venus, there would be no chance of a landing. That planet, with its thick, hot atmosphere, has a surface temperature of 470° C. everywhere, on the night side as well as the day side. Any exploration of the surface would have to be carried out by unmanned probes launched from the space vessel, which would itself remain in orbit about Venus and which would eventually have to launch itself into another six-month journey back to Earth.

Since exploration of Venus's surface would have to be carried out by unmanned probe, that probe might as well travel all the way from Earth. The benefits achieved by having the probe launched from, and signals received by, a manned mother ship would scarcely justify the traumatic experience of some twelve continuous months in space.

Mercury, the planet nearest to the Sun, is farther from us than Venus is, being never closer to us than 80 million kilometers. The elliptical orbit that would take astronauts there is not enormously longer than that to Venus, and astronauts could at least land on Mercury.

Mercury has only a trace atmosphere, incapable of conserving heat. In the course of its slow rotation (fifty-seven Earth days), the Mercurian surface can heat up in spots to temperatures higher than that of Venus, but in the course of the Mercurian night, the temperature of the surface would drop rapidly and present no problem. Astronauts could explore a given area for a month between sunset and sunrise.

Although the total elapsed time represented by a flight to Mercury and back would be greater than that to Venus and back, the former would be broken midway by a planetary landing, a chance to uncramp from the confined quarters of the space vessel, and that would make the trip to Mercury less psychologically difficult than the one to Venus.

The flight to Mercury would, however, carry the astronauts and their ships to some 65 million kilometers from the Sun, even when that small planet is at its greatest distance from the solar furnace. Solar radiation would be over four times as concentrated at that distance as it is in the neighborhood of the Earth. For what might be gained in a long, manned voyage to Mercury, the price paid in risking the effects of the greater radiation may prove too high.

Since voyages in the direction of the Sun offer no suitable target, what about voyages away from the Sun?

The nearest planet to Earth in the other direction is, of course, Mars. It is, at its closest, some 58 million kilometers away, so that while it is farther than Venus, it is closer to us than Mercury is, and traveling Mars-ward means steady progress in the direction of decreasing intensities of solar radiation. Furthermore, it is a cool world (cooler than Earth) and can be explored in comfort for indefinite periods whether the Sun is in the sky or not. There is no fearsome Sunrise acting as a deadline, as there would be on Mercury.

In almost every respect, Mars is more interesting than the Moon. Mars is the larger world; it has an atmosphere, albeit a thin one; it has some water, albeit very little; it has polar icecaps and seasonal changes; it has an active geology that produces volcanoes; it has the possibility of long-range cyclical changes that may produce a milder Mars, at intervals, with a thicker atmosphere and free liquid water, and consequently the possibility of a native life.

But, in one respect, Mars is a more difficult target than the Moon is. It is 150 times as far away as the Moon is, even at the point of closest approach.

The round trip, to Mars and back, would take a year and a half at the very least. Even though that will be broken, for a shorter or longer time, by a landing on a planet which, next to Earth itself, is the most comfortable in the Solar System, the task would surely stretch human endurance to the limit.

And beyond Mars? To reach the larger asteroids, or the satellites of the giant outer planets, Jupiter, Saturn, Uranus, and Neptune, would take years and even decades. Manned voyages of such lengths do not seem practical at the moment.

In addition to the Moon, then, we are left with only Mars as a sizable target, and that only as a borderline possibility. There would also be the smaller bodies that venture within the Martian orbit. These would include the occasional asteroid such as Eros or Icarus, and the occasional comet such as Encke's comet. For completeness, we might also mention the two tiny satellites of Mars, Phobos and Deimos.

All in all, the possibilities beyond the Moon offer us a small and disappointing range of targets. It would seem that in the present state of the art, the Moon is not only the threshold of space but, as far as manned exploration is concerned, virtually all of it.

Yet surely it is a mistake to limit one's self to some existing state of the art, without considering the possibility of advance. To have done so in 1900 would have made the Moon unreachable to mankind; to have done so in 900 would have made North America unreachable to Europeans.

The difficulty in moving to the worlds beyond the Moon is one of distance, and of the time it takes to cover that distance. Years and decades in space in the kind of vessels we can now build is simply not in the cards.

Can distance be eliminated altogether by removing the necessity of a slow acceleration consistent with what the human body can endure, and therefore the achievement of high speeds in a very short interval of time? If we could combine this with the attainment of speeds far in excess of the speed of light, not only might the outer regions of the Solar System be reached in a reasonably short time, but the stars themselves might fall within our grasp.

(As long as the speed of light remains a limit, even relatively near stars require round trips lasting decades, while moderately distant stars require centuries of travel. To cross our galaxy would take some hundreds of thousands of years, and to reach any object outside our galaxy and its immediate neighbors we would find ourselves dealing with journeys of millions and billions of years in length.)

How can we possibly bypass the speed of light, though, when that is the ineluctable limit set by the theory of relativity? Some physicists, however, have postulated particles of matter that have the property of traveling only at speeds *greater* than that of light, and show that this, too, is consistent with relativity. Either always slower or always greater, but never both. The faster-than-light particles are called "tachyons," from a Greek word for "fast."

We might imagine, then, that every subatomic particle making up a ship and its contents might be translated into the corresponding tachyons instantaneously. These would travel at enormous speeds for some desired period of time in the form of "tachyonic ships" carrying a "tachyonic crew." They would at a given moment change back, again instantaneously, into ordinary particles. In less than a second, a ship might, in theory, travel any distance—from end to end of the Universe perhaps.

Tachyons, however, have been only postulated, so far; they have not been detected. Even if they existed, the necessity of changing *every* normal particle into tachyons, and then reversing the change in virtually perfect simultaneity, would be a most harsh requirement. (If some particles make the change a fraction of a second out of tune with the others, the tachyonic ship and crew may end up spread over light-years of space.) Finally, even if the changes could be made, the task of directing and controlling tachyonic flight would represent a most formidable engineering problem.

All told, while we can speculate about tachyonic flight, it is not reasonable to consider it seriously. It can come about only as a result of a fortunate and as-yet-unlooked-

367

for series of scientific breakthroughs, which may very well never happen.

Even if the speed-of-light limit is never surpassed, it might still be possible to eliminate the *sensation* of time passage. We might imagine a ship traveling through interplanetary and even interstellar space at quite a low velocity but with its astronauts in a frozen state of suspended animation and the ship itself under strictly automatic control. On approaching some destination, the astronauts could be automatically thawed and roused.

This, again, implies the development of techniques that, at present, we have no right to assume as an inevitable development. Relatively simple life forms can be frozen and revived, and it seems quite likely that living organisms frozen quickly enough and brought to low enough temperatures (say, those of liquid air) might remain suspended indefinitely in a state that can be brought back to life.

Nevertheless, the difficulties of freezing a living object as large as a man sufficiently quickly to produce no life-destroying damage in the process are enormous. No warm-blooded creature has as yet been thoroughly frozen and then restored to life, and we cannot be confident that this will ever be successful. We must therefore allow frozen flight to remain, along with tachyonic flight, a subject for far-out speculation only, and not for serious anticipation.

There is, however, another way of eliminating the sensation of time passage, while allowing the astronauts to retain full consciousness—a method that is known to be possible in the light of modern knowledge and that requires no unforeseen breakthroughs.

The theory of relativity makes it quite clear that as the speed of a vessel increases relative to the Universe generally, the time rate that is experienced slows, reaching zero at the speed of light.

If a ship's velocity is raised to nearly the speed of light, then, to the astronauts on board, a voyage that would ordinarily be experienced as having endured centuries would seem to have endured, to the vastly slowed time sense, only years or even weeks. Under such circumstances, the astronauts, by edging closer and closer to the speed of light, could cover longer and longer distances in a given time interval and would, in effect, accomplish what a tachyonic ship would.

There are, however, catches to travel by time dilatation, as compared to travel by tachyons, that remove some of the bloom from the former.

1) If we could turn ordinary subatomic particles to tachyons, that might, in theory, be done virtually instantaneously. To attain near-light velocities for ordinary particles which, however, will remain as such, there must be acceleration first and then, as the destination nears, deceleration. The rate of neither acceleration nor deceleration can be very great, in view of the fragile structure of the human body, so the task of getting up to high speed and getting down from it is very time-consuming (and very energy-consuming as well). Interstellar travel will therefore always take some irreducible and uncomfortably long time, despite the time-dilatation effect.

2) Once a very high velocity is achieved, there is a problem concerning the interstellar medium itself. We consider it a vacuum, and it is certainly a better vacuum than anything we can make here on Earth. At very high speeds, however, a ship will pass so many of the thinly spread atoms that exist even in interstellar space, that what seems a vacuum at ordinary speeds will become a highly resisting medium at high speeds, and that may set a practical limit to how fast one can go. That limit may be sufficiently far from the speed of light to reduce the time-dilatation effect considerably.

3) Even if the time-and-energy problem is dismissed and if we assume that near-light velocities can be achieved without interference from the interstellar medium, the fact remains that time dilatation affects only the speeding astronauts and not the rest of the Universe generally. The space vessel and its crew may indeed be able to reach the Andromeda galaxy in, let us say, a year, explore some part of it for a year, and return in another year—but on Earth, lacking the time-dilatation effect, four and a half million years would have passed. (This sort of thing would not happen in the case of tachyonic travel.)

It is doubtful if astronauts would volunteer for a space voyage if they knew that they would never return to any Earth they could recognize. And even if some would indeed be willing to wash their hands of Earth forever (perhaps with a muttered "Good riddance!"), it is considerably more doubtful that Earth would be willing to invest in an exploring mission of which no human being then alive (and possibly no human being who would ever live) would see the results. The question "What's in it for me?" in connection with the expense and effort of the voyage would have as its answer "Nothing!" and so it would not be done. It might be, then, that only the very nearest stars would be within reach, no matter how high a velocity could theoretically be reached and no matter how slow time could be made to progress for the travelers themselves.

There seems to be only the alternative of accepting the limitations of both time and distance and of making the best of it. One can point a ship at the stars, let it gain some moderate velocity, and then let it coast indefinitely. To reach even the nearer stars under such circumstances might take thousands of years as judged by the astronauts themselves as well as by the stay-at-homes on Earth. Generations of astronauts would have to be born, live out their lives, and die, in the course of so mighty a voyage.

Naturally, this could not be done on board any spaceship of the kind with which we are now familiar. There is no life-support system we can now build that would last for years, let alone millenniums, and it isn't conceivable that astronauts would be willing to pass their lives in constricted quarters and have children who would then have to pass *their* entire lives in those same constricted quarters.

It would be necessary to have a ship large enough to represent a world—a small one, perhaps, but one sufficiently elaborate for each member of the crew to find his immediate surroundings worldlike. The dimensions of the ship would have to be measured in tens of kilometers and the population on board would have to number in the hundreds of thousands.

Let us call such large vessels, capable of supporting a large population indefinitely by on-board agriculture, animal husbandry, and industry, a "starship."

The concept of the starship requires no unforeseen breakthroughs and seems to possess no hidden technological catches—but there are psychological problems:

1) Merely building a starship would require some tens of billions of present-day dollars, some tens of billions of tons of materials of all kinds, including a vast quantity of energy. Even if such a project is feasible from the technological standpoint, would the population of Earth be willing to devote the effort and the resources necessary for the purpose?

2) This is especially so since, as in the case of a time-dilatation ship, there is no chance of a starship returning to Earth except after a huge lapse of time. Again, the population of Earth would scarcely be willing to invest so much when there will be no tangible return of any kind in the foreseeable future.

It would seem, then, that even looking hopefully into a future of advanced technology, we may not make many advances in the exploration of space. It almost seems that no matter in what direction we look we end with the Moon being both beginning and end; no human exploration seems possible beyond the Moon—now or ever.

But can this be so? The thought of so-restricting a limitation is abhorrent to any romantic (and all explorers must be romantics), and we must therefore look further and search avidly for any loopholes that might brighten the bleak picture I have drawn so far.

Throughout the discussion so far, I have made the assumption that the home base for space exploration will be the Earth; that it will be from Earth that our explorations will start. Might it not be, though, that if we can broaden our base, the situation with respect to the further exploration of space might change—and for the better? After all, it could hardly change for the worse.

And surely our base will broaden. If our civilization survives,[1] it will inevitably broaden. However narrow our capacities for exploring space may be, they are broad enough to include the Moon. We have already reached the Moon and returned safely six different times. If we develop the space shuttle and establish space stations more advanced and versatile than Skylab, we can initiate flights to the Moon that will cost far less than those of the Apollo program and be capable of accomplishing more. Given a reasonable advance in technology, we can establish a permanent, ecologically independent colony on the Moon.

The Moon seems a harsh and forbidding world by Earth standards, but most of those factors that make it seem forbidding are properties of the surface only. It is the surface that is subjected to a two-week siege of Sunlight that brings the temperature, in spots, to the boiling point of water—followed by a two-week absence of Sun that lowers the temperature from boiling down to halfway toward absolute zero. It is the surface that is subjected to meteorite bombardment, to the harsh radiation of the Sun, and to the cosmic radiation from points beyond. It is the surface that lacks air and water, and could not hold on to it (thanks to a surface gravity only one sixth that of Earth) even if air and water were somehow supplied and placed upon that surface.

Once caverns are excavated beneath that surface, though, and several meters of lunar soil are placed between its deadly properties and humanity, a comfortable world could be built. Beneath the surface, temperature would be equable and uniform and there would be neither day nor night, neither summer nor winter. With artificial lighting, time could be organized for human convenience.

There would be no danger from meteorites[2] or from radiation either.

To begin with, of course, there would have to be a large capital investment from Earth. Human beings, plants and animals, machinery, energy would all have to be supplied by Earth at first but over a period of time, not all at once. Little by little, moreover, the lunar colonists would begin to draw upon the Moon itself for resources.

Energy, for instance, would be easily available on the Moon. Any given spot on the Moon experiences Sunlight for two weeks at a time, with no interference ever from clouds or dust (points on the lunar equator experience the maximum quantity). Solar cells could be spread out over almost unlimited areas. There is no weather to interfere, no vandals to fear, no native life forms to displace, no native ecology to upset.

The energy so obtained could be used to power the chemical processes that will release, from the molecules of the lunar soil, the various elements that can be combined to form other molecules. The elements themselves, and the molecular compounds, whether native or synthetic, will be used in construction, in the building of machinery, in the million and one uses already worked out on Earth.

Energy in the form of an electric current could split water into hydrogen and oxygen. The oxygen could be used in building an atmosphere, the hydrogen in chemical syntheses. Unicellular organisms, growing rapidly under the encouragement of human-expired carbon dioxide, artificial light, chemical fertilizers, properly treated human waste, could form protein at breakneck speed with almost no waste. Some of them could renew the atmosphere, and all would remove waste and serve as a food supply.

Eventually, some forms of animal life could be raised on the Moon, and multicellular plant life cultivated there. A reasonably normal human dietary standard might be established.

A major flaw in this picture of a flourishing lunar colony has been revealed as a result of the manned exploration of the satellite. The rocks brought back to Earth show that the Moon's crust is low in the content of the more volatile elements—those that form compounds that are low-melting. Presumably, the Moon went through more or less extended periods at elevated temperatures and lost volatiles by vaporization.

In particular, water is absent. Judging by the nature of the lunar rocks that have been studied, it would seem that the lunar crust is everywhere thoroughly dry.

This may turn out to be an overly pessimistic conclusion, but even if we allow a totally dry Moon, a lunar colony need not be ruled out. Earth has, if anything, an oversupply of water, for if a slight rise in temperature were to melt the planetary icecaps, the coastal rims of the continents would be drowned, producing unimaginable disaster.

It would be no great sacrifice to offer the lunar colony some tens of thousands of tons of water; or merely hydrogen, which could be turned to water by combination with oxygen from the lunar crust.

Since the lunar colony would carefully recycle everything as efficiently as possible (as, otherwise, survival would be questionable), not much in the way of new material would have to be introduced. Small capital investments of hydrogen, for instance, would last a long time. Additional supplies would be required more to sustain colony growth, perhaps, than to replace cycling losses.

To be sure, it may be that Earth, overaware of its own needs, would choose to grant only grudgingly and sparingly the volatiles required by the lunar colony. It may also be that the lunar colony, oversensitive to its own dependence on the home world, would seek some source of volatiles other than Earth.

It might seem, at first, that there is no other practical source within reach for volatiles (not only for water but for compounds of such elements as carbon and nitrogen, in which the Moon is relatively lacking and which are particularly important to life).

Venus possesses volatiles, but these are present in its atmosphere only in gaseous form and would therefore be difficult to gather in quantity. Mars has sizable icecaps containing frozen water and frozen carbon dioxide, but Mars offers the next possible base for the expansion of humanity and there would be serious ethical questions as to whether that planet's limited store of volatiles ought to be rifled.

Where else could the lunar colonists turn? In the outer Solar System, where the intensity of solar radiation has always been low, the real supply of volatiles is to be found. Some of the larger satellites may possess significant amounts. Ganymede and Callisto, which circle Jupiter, and Titan, which circles Saturn, are thought to be rich in volatiles.

375

The outer Solar System is at a great distance, however, and the lunar colonists, at least in the early years of their existence, would surely feel the need of something closer.

Fortunately, there is an alternative. Not everything that exists in the outer Solar System remains there permanently. There are some objects in the Solar System that have highly elongated orbits—the comets. At the far end of their orbits, they are in the outer Solar System, far beyond even the farthest planet in some cases. At the other end of their orbits, they pass through the inner Solar System.

The comets, originating (it is thought) in the far reaches of space, a light-year or more from the Sun, consist, to begin with, of "ices," that is, of frozen volatiles. (Observations of comet Kohoutek from Skylab in 1974 offered the final confirmation of this.)

If comets remained in the far-off cloud in which they formed, they would remain frozen and intact indefinitely. Every once in a while, however, a comet slows in its orbit because of the pull of some distant star and drops in toward the Sun.

Comets lose some of their volatiles at each entry into the inner Solar System and with each whirl about the Sun. Those volatiles boil off to form a foggy "coma," which is then driven outward by the solar wind into a long, filmy tail. Comets that take up relatively short orbits (because of the gravitational perturbations of the planets they pass) approach the Sun at frequent intervals, losing their volatiles in time and leaving behind a rocky core, or nothing more than a thin cloud of involatile dust. There are always comets, however, which have made, as yet, few approaches, and are still rich in volatiles.

It may be that, in time, the lunar colonists, having struggled along with what skimpy supplies of volatiles they have squeezed out of a reluctant Earth, will have developed the techniques for trapping such comets.

Moon-based telescopes could detect such comets far out in space, even before they had reached Jupiter's orbit. (To be spotted far off is itself the sign of a large, new comet, possibly very rich in volatiles.) The comet's orbit would be plotted, and in the months it takes to approach and enter the inner Solar System, the lunar colonists would place a ship at some rendezvous point in space.

A landing would be made on the comet, which might be no more than a few kilometers across the solid core. Rockets, appropriately placed on the comet, which make use of the substance of the comet itself as the material to be turned into exhaust once heated, or possibly the use of some advanced nuclear drive, would force the comet out of its orbit.

Little by little, the comet's motion would curve in such a way as to bring it slowly closer to the Moon, then into orbit about the Moon, then spiraling down to the Moon's surface. Finally, it might be brought down within the southern lip of a north-polar crater, for instance, where, in the eternal shadow of the crater wall, it would remain permanently frozen.

The whole process would be like that of hooking, maneuvering, and landing a gigantic fish. Such a "beached" comet, with cubic kilometers of volatiles, would easily make the lunar colonists independent of further supplies for decades. And long before the supply would be consumed, another comet might come into view.

It is not necessary to discuss here the uses and values of a lunar colony. These are many, for it could serve as a base for research that cannot be performed on Earth; a place where industrial processes could take place in super-Earthly fashion by making use of un-Earthly conditions of low temperature, high vacuum, intense solar radiation, and so on; a place that might serve as a first attempt at preparing a thoroughly engineered abode for human beings and as a working model for Earth of a low-birth-rate, resources-conserving-and-recycling economy, by which alone our home planet can survive.

All this could eventually make the lunar colony hugely profitable to Earth, but never mind that. What I shall discuss here is how the existence of a flourishing, prosperous, vigorously growing lunar colony would affect the program for the further exploration of space.

The prospects would surely change dramatically once the lunar colony was a going concern, for the interaction of lunar colonists and space is bound to be entirely different from that of Earthpeople and space.

1) Space flight is an exotic matter to the people of Earth, something that would take them away from the world on which they live and on which they have developed over a period of billions of years. Space flight would, on the other hand, be of the very essence of life to the lunar colonist, whose world would have been populated as a result of space flight and which would be growing and prospering only through volatiles obtained by space flight (whether from Earth or elsewhere). Where Earthpeople might, on the whole, hesitate to venture into space, that would come as second nature to the lunar colonist.

2) The conditions of space flight represent an extreme changeabout to us of Earth. We are accustomed to clinging to the outer surface of a very large world; to a cycling of food, air, and water through so vast a system that one is scarcely aware of it; and to all the accouterments of such a world—the blue sky and green land, the sound of birds, the

smell of flowers, and all the rest. Getting into a spaceship would mean inhabiting the inside of a very small world—the cycling of food, air, and water in so tight a fashion as to prove an ever-present fact forcing itself on the consciousness of the crew. And, of course, there would be none of the pleasant side-effect properties of Earth.

For lunar colonists, however, the change would not be at all extreme. They would live, in any case, on the inside of a world that is, to be sure, a fairly large one taken in all, but in which the cavern volume forming the effective habitat would be small. The colonists would be accustomed to close cycling and to a thoroughly engineered environment. They would have none of the accouterments of Earth.

In short, the lunar colonists, in undertaking a space flight, move from one spaceship to another, very similar, but somewhat smaller spaceship.

This does not make the space flights to some specific destination less long or less dangerous, but it does enormously lessen the psychological difficulties. A crew of lunar colonists could undoubtedly endure the restricted quarters of a spaceship for a year or more far more stoically and efficiently than an Earth crew would.

3) Since the surface gravity of the Moon is only one sixth that of the Earth and since there is no atmosphere on the Moon, a spaceship take-off from the Moon requires far less energy than it does from Earth, and runs no risk of overheating through friction. What is more, most of the destinations for space flight in our Solar System are much more Moonlike in character than Earthlike—gravitationally and in every other way. Lunar colonists would feel more at home on asteroids and satellites than ever Earthpeople would.

4) The fact that the lunar colonist would be accustomed to a gravity only one sixth that of Earth is important. De-

veloping under a low gravity, he would be likely to possess slimmer and more delicate bones and muscles and might be more dextrous and delicate in the manipulation of machinery and controls. It may also be that he would be less likely to suffer loss of muscle tone and bone calcium under prolonged subjection to zero-gravity conditions, losses that have affected Earthpeople so subjected.[3]

It may well be, then, that lunar colonists could make long trips under zero-gravity conditions, which to Earthpeople would be impossible.

Suppose, though, that lunar colonists are adapted to low gravity but cannot withstand zero-gravity conditions any better than Earthpeople can. It may then be necessary to produce a gravity substitute, in the form of the centrifugal effect, by placing a spin on the vessel.

The effect depends for its intensity on the rate of spin and on the distance of an object from the center of rotation. If the lunar colonists are well adapted to low gravity, any ship they are on will need a rate of spin only one sixth as rapid as that of an equivalent ship with Earthmen aboard —thus reducing the engineering problems. (To be sure, lunar colonists would probably tolerate lower accelerations than Earthpeople would, which would lengthen flights— one point against them.)

If we take all this into consideration, it may well prove, on balance, that although Earthpeople may reach Mars as a tour de force before a lunar colony is brought into full, flourishing existence, it will be only after that colony is mature and only after lunar colonists, born and bred on the Moon, are launched into space, that trips to Mars and back become a routine affair. It is only with the voyages of the lunar colonists, facing a year or two in space with equanimity, that the Martian surface will feel the footsteps of human beings in wholesale quantities; that the planet might be explored; and that the first engineering steps might be taken to make it safe for mankind.

Of course, the lunar colonists might not, themselves, colonize Mars. The Martian surface gravity is two fifths that of Earth, but still two and a half times that of the Moon. This raises a problem, for one can perhaps travel from a life-long accustomed strong gravitational pull to an unaccustomed weak one with no pain and with easy adaptability, but the reverse can scarcely be easy.

The lunar colonists, then, may find it uncomfortable to remain very long on the Martian surface. They may have to establish their base on Mars's tiny and nearby satellite, Phobos, and explore and engineer the planet in shifts.[4]

Once lunar colonists have done their work, however, people from Earth can travel to the Moon, and then be brought to Mars as colonists in modified lunar ships (larger and of faster spin) manned by lunar colonists. Mars may then form a third world inhabited by human beings.

Once it is reached, Mars may prove far easier to colonize than the Moon would be. The Martian gravity is closer to that of Earth; it has a thin atmosphere, though an unbreathable one; and it has an Earthlike rotation, of just over twenty-four hours. Its greater distance from the Sun keeps the surface temperature cool and minimizes the danger of solar radiation. (Since it is closer to the asteroid belt, Mars may suffer from a somewhat greater meteorite bombardment intensity, however.)

Most of all, Mars has an ample supply of volatiles of its own. Certainly, the quantity is very small on an Earth scale, but after all, the supply will not have to meet the needs of a planetary ecology, but only those of a limited Earth colony.[5]

In some ways, the Martian colonists will enjoy the advantages of the lunar colonists as far as space flight is concerned. Like the lunar colonists, the Martian colonists will be living in an engineered underground world and will not feel a strong psychological wrench in transferring to a smaller engineered underground world. What's more, the Martian colonists will be closer to the vast, unexplored reaches of the outer Solar System.

Still, Mars, with its atmosphere and its greater gravitational field, is a more difficult base from which to launch spaceships, and its colonists will find it harder to adapt to zero gravity. On the whole, then, the lunar colonists may retain their virtual monopoly on major space flight.

In fact, we might envisage a twenty-first century in which Earth and Mars are occupied by relatively immobile and "landlocked" races of *Homo sapiens*, while the lunar colonists will launch themselves fearlessly into the vast ocean of space, playing the role, over a greatly enlarged sphere of action, of the Phoenician and Polynesian voyagers of the past.

The voyages of the lunar colonists will undoubtedly be important even if they land nowhere, for they can test the properties of the space surrounding the major planets and at varying distances from the Sun. Still, they might easily establish bases in the outer Solar System. Except for the giant outer planets themselves (which are likely to remain unapproachable by mankind in the foreseeable future), all bodies in the outer Solar System have surface gravities in the lunar range or less and would be easily occupied by the lunar colonists.

The larger asteroids are obvious targets for bases where a spaceship on a multi-year flight may stop for "rest and relaxation."

The same might be said of the satellites of the outer planets, though here difficulties might arise in connection with the magnetic fields of the primaries. The large, Galilean satellites of Jupiter are within the enormous Jovian magnetic field, and human penetration of that field would raise enormous questions as to shielding. Callisto, the farthest of the Galilean satellites, may be relatively safe, and certainly the small, outer satellites of Jupiter (eight of these are known at the moment) would be.

We might visualize numerous bases established throughout the Solar System in time, one even as far out as distant Pluto, where the initial purpose of the colony might be to devote itself to the study of the stars in a sky in which the Sun is reduced to nothing more than an extraordinarily bright star itself. Pluto, with a surface gravity about that of Mars, may have fearfully low temperatures, but, given energy, it is easy to heat low-temperature caverns. While solar energy would be unattainable on Pluto, by the time it is reached mankind ought to have useful nuclear-fusion reactors and the planet should have a supply of hydrogen to use as fuel.

The Plutonian colonists may represent a third relatively immobile race of *Homo sapiens*, one that would be fearfully isolated[6] and would be utterly out of reach of other human beings except by way of radio communication (itself not an easy task over enormous distance, where it would take hours for a speed-of-light message to go and an answer to return). There might be an occasional visit by a lunar ship, however, or perhaps, more likely, by a Callistan or Titanian ship.

And in ships better shielded against radiation than any we can build now, the lunar colonists may head inward toward the Sun, studying the properties in the neighborhood of Venus and of our luminary itself, and effecting an actual landing on Mercury.

If all this comes to pass, then, we will find the Moon to be indeed a threshold to space. While merely reaching the

Moon does us little good, colonizing the satellite would open the entire Solar System to human colonization and development. The result of that would be to extend greatly the knowledge that will be available to human beings; greatly multiply the variability of human culture; and greatly increase human security in case of disaster, since no longer would all human beings be found on a single world.

If the Moon is the threshold to the exploration and colonization of the entire Solar System, will it help us move beyond the Solar System to the far greater Universe of the stars?

The various advances in technology that might make the exploration of the stars possible—faster-than-light drives, near-light velocities, long-term freezing—have been considered and found wanting with Earth itself as an exploration base. They are still to be found wanting and their uncertainties and flaws are not in the least ameliorated by the fact that mankind may be on the Moon, or on any other world of the Solar System. With the entire Solar System triumphantly penetrated, exploited, and colonized, the stars will still remain unreachable, in all likelihood, by such techniques.

But what of the last alternative? What of the construction and use of a starship? Does that become in any way more reasonable and practical because there are colonists on the Moon?

In some ways, yes.

The construction of a starship on the Moon would be very much equivalent to the construction of a cavern extension under the lunar surface. The lunar colonists would have the necessary skills and experience and could build a starship more efficiently than Earthpeople could.

What's more, the materials for the starship would be drawn from the Moon itself. Since the lunar surface is dead and since the lunar colonists will, for a long time, take up only a small portion of that lunar surface, there will probably be no sense of loss in starship construction on the Moon.

Earthpeople are bound to feel that every ton of aluminum, steel, concrete, or glass taken from Earth and used for a starship is a ton abstracted from possible use by Earthpeople and a ton whose removal diminishes the strength of an already enfeebled terrestrial ecology. The lunar colonists, on the other hand, are likely to shrug off the ton used as a ton that would otherwise lie in the ground uselessly for an indefinite period of time.

Once built, a starship manned by lunar colonists would offer an environment even more like the lunar colony itself than an ordinary space vessel would be, and the starship could be launched from the Moon with far less energy expenditure than what would be necessary on Earth.

Yet even so, one can imagine disadvantages. For one thing, there remains the difficulty involved in leaving home *forever*. Lunar colonists who might be willing to undertake voyages lasting for years may balk at voyages lasting forever. Worse yet, the lunar colonists remaining at home may balk at building a ship that will never return. The inhibiting effect of "Good-bye forever!" might well be as effective on the Moon as on the Earth in stalling the starship project.

There is also, perhaps, a physiological difficulty. A starship, we might argue, ought to have some gravitational effect. Even if men adapted to low gravity could endure zero gravity over space flights with durations of several years, it might be that the actual birth of children under zero gravity, and their growth and development under zero gravity, would prove impossible.

It would be a difficult thing to check, but the testing of

birth and infant development in an orbiting satellite might be attempted, and might turn out to be feasible. Zero-gravity children might develop a whole new set of conditioned reflexes and learned responses that would make it second nature for them to move and to handle objects under zero gravity, so that they will do it as easily and deftly as we work under normal gravity.

Even if that were so, however, a starship under zero gravity would not be advisable. It would make of the human population of the ship permanent prisoners. They might well be unable to step out upon any world larger than a moderately small asteroid. It might be that any perceptible gravitational pull, one of only a pound or two, might be frightening, nauseating—even physiologically unendurable.

Surely if a starship is a true exploring device, its crew should be able to step out on a reasonably large world and not needlessly limit their range to bodies of insignificant size only.

A large ship, set spinning at a moderate rate about its long axis, would force objects toward the inner surface of its shell and away from the axis in all directions. Wherever one stands in such a ship, the direction toward the axis would be up and the direction toward the shell and open space would be down (rather the reverse of the situation on a sizable world).

This, however, might introduce a difficulty.

In any world large enough to possess an appreciable gravity, the gravitational intensity decreases with distance from the center, but that distance is already so large at the surface, that it takes a considerable further increase to produce a noticeable decrease in gravity. One would have to rise to a height of thirty-five kilometers above sea level on Earth (four times the height of Mount Everest) before gravitational intensity would drop by 1 per cent. It would require a height of nineteen kilometers above the average lunar surface level to cause the Moon's gravitational intensity to drop by 1 per cent.

On such worlds, the gravitational intensity is roughly constant when one is on or near the surface, and all the sensations, reflexes, and responses are conditioned to a constant gravitational intensity.

A cylindrical starship, three kilometers across the short axis, let us say, and rotating about its long axis, would produce a sensation of gravitational attraction toward the shell that would decrease linearly as one moved toward the axis, reaching zero at the axis. If one moved fifteen meters away from the outer shell toward the axis, the sense of gravitational pull would drop 1 per cent in intensity.

Moving through the ship would surely involve relatively large changes in gravitational effect, which would itself produce physiological, psychological, and engineering difficulties.

Undoubtedly, if we are liberal enough to suppose that the human body can adapt itself to low gravity and even zero gravity, we might go a step farther and assume it can adapt itself to variable gravity as well—but it nevertheless adds one more difficulty to the "Good-bye forever" syndrome.

For these several reasons, then, it might be that even with a lunar colony in successful operation and with a Solar System fully developed, the exploration of space beyond the Solar System might be stymied.

It may be, though, that we are overlooking another way of expanding the human base for exploration, one that could be perfectly feasible and that might then make exploration beyond the Solar System a completely practical matter.

Suppose that the lunar material and lunar expertise were used to build a starship but that the starship was *not* launched on its endless voyage into space but was kept in the Solar System.

Gerard K. O'Neill of Princeton University suggested, in 1974, that the equivalent of what I have called starships could be built in the near future and made into orbiting space colonies. In his view, the lunar colony, which has always been taken for granted as the first step in the expansion of humanity beyond Earth, could well be overleaped. People would land on the Moon, yes, but not to colonize it —only to convert it into an enormous quarry. Material from the lunar surface could be towed out into space, where it could be smelted and converted into steel, aluminum, and glass, then put together as a large container for soil (also from the Moon) together with machinery, plants, animals, people, and liquid hydrogen—all of which would be initially from Earth.

O'Neill visualizes the first space colonies as being placed in the Trojan position with respect to Earth and Moon: the three bodies—Earth, Moon, and space-colony cluster— would form the vertices of an equilateral triangle. The colonies would be 380,000 kilometers from Earth and 380,000 kilometers from the Moon. The colonies would move about the Earth in the Moon's orbit, lagging behind the Moon by 60° or moving ahead of it by 60°.

There might be many such colonies of varying size. The smaller ones, with lengths of a kilometer or so and widths of a tenth of a kilometer, might house a population of thousands, while those thirty kilometers long and three kilometers wide might house millions.

It is difficult to judge whether it is more feasible, from a political, economic, and psychologic standpoint (not technologic), to build the space stations directly by Earth labor and use the Moon as a quarry, or to wait for a functioning lunar colony to take the matter in hand.

It might turn out that the intermediate production of a lunar colony would work better. The Moon is a world, and mankind is presently accustomed to worlds. Moon colonies can begin very small indeed, whereas a space colony would probably have to be fairly large to begin with, in order to supply a sufficient thickness of material to protect the population against cosmic-ray bombardment. (The lunar crust would supply that protection for even the smallest lunar colony.)

This means that although a functioning lunar colony might cost more than a space colony, or even several space colonies, the expense might be spread out over a long time interval and be less painful, all told.

Once the lunar colony is flourishing, it would be populated by human beings far more space-minded than our Earth population and far readier, psychologically, to build space colonies.

Of course, whether built directly by Earth labor or eventually by lunar-colonist labor, the end result would be the same. The ships that would explore, exploit, and colonize the Solar System could have a lunar-colonist crew or they could have a space-colonist crew.

The point of discussion here, however, is whether the existence of space colonies would make *stellar* exploration more feasible than the existence of a lunar colony (and of colonies on other worlds of the Solar System) would.

Consider that a space colony is, in actual fact, the starships we have been discussing, except that the space colonies lack certain of the properties that place difficulties in the way of exploration beyond the Solar System.

The Frontiers of Knowledge

First, the initial space colonies, located at the Trojan points of the Earth-Moon system, would remain within the Solar System and, indeed, very close to home and to the oldest and most advanced populated worlds in existence. They would not be moving off into infinite space forever, so that the inhibiting factor of "Good-bye forever" would not exist. The space colonists would not be abashed and distressed by the prospect of lifelong isolation for themselves and their descendants, and the lunar colonists would not have to watch their entire investment disappear forever. This means that the way of life of a starship could be worked out in comfort and security and with full support from older worlds.

For instance, there would still be the matter of adjusting to the markedly variable gravitational effects, but this could now be done without the added difficulties of breaking the umbilical cord. Those who could not manage to adjust to a variable gravitational intensity could return to the Moon for rest and rehabilitation before trying again—or could give it up forever. It may be that the very knowledge that one could give up if one wished would make it easier not to give up at all.

In fact, we might suppose that variable gravity, once adapted to, would have its own joys. The mountains built along the inner walls of the far ends of a large space colony would offer a new kind of mountain climbing, in which the higher one climbed, the easier it would be to climb still higher. There might be the joys of personal flying, with wings strapped to one's arms, if one rose at a level near enough to the axis. There could be fun with zero gravity at the axis.

In short, the space colonies would offer a way of developing, in stages and with security, a population thoroughly adapted to the starship way of life.

Then, too, these space colonies would differ from the

390·

inhabited natural worlds with respect to mobility. The Earth, the Moon, the various other inhabited bodies of the Solar System, would all be too massive to respond easily to thrusts designed to move them far out of their orbit. It would take an extraordinary and quite prohibitive quantity of energy to move them out, largely because of the over-whelming fraction of their bodies that has no connection with life except to serve as inertia-originating mass.

A space colony, which has insignificant mass by world standards, could be moved out of orbit with a comparatively small output of energy. It *could* take off for the stars, if it wanted to, and it would have a new and vast advantage over even the lunar colonists when it came to making the journey.

A lunar colonist, in transferring from his home on the Moon to a spaceship, is making a transfer from a larger place to a smaller place of the same sort, and that gives him an advantage over the inhabitants of Earth. Nevertheless, he is still leaving home. A space colonist, however, has a further advantage, since he makes no transfer at all! His *home* moves. He can travel to the end of the Universe without budging from his own familiar surroundings.

The fact that this is so may still not make it easy for him to leave the Solar System. Although a space colonist has broken away from Earth and Moon alike—from all worlds —he has not broken away from the Sun. If the space colonies are indeed built as part of the Earth-Moon system to begin with, then all will consider the Sun—an object with the appearance in the sky of *our* Sun—to be part of home.

Something else is still needed to make stellar exploration feasible, and that something else might come about quite inevitably.

The space colonies would be bound to feel a kind of

population pressure. The success of one space colony will make it that much more pressing or urgent—or, in any case, desirable—to build another. With room made available for millions of people in space colonies, the human population (which might have been stabilizing its numbers or even bringing them down) might expand again. Populations, after all, grow to fill the available room and more; this has always happened.

To have a number of space colonies at each Trojan point would be pleasurable. Space colonists could travel from one to another with virtually no expenditure of energy, and each space colony would seem novel and interesting to the visiting tourists from neighbor colonies.

With multiplication, however, there will inevitably come to be too many colonies. Orbital perturbations will limit how many space colonies can safely be crowded into each Trojan point, the lunar colonists may balk at having to spend more and more of their time, effort, and resources on endless production of space colonies, and the space colonies themselves will feel unpleasantly crowded.

The push will be on for the space colonies to move outward in search of empty space and of new resources for the building of daughter colonies of their own.

The logical way out would be to move to the asteroid belt, where some hundred thousand asteroids can supply the necessary materials (including volatile matter on some of the larger ones, perhaps). The total mass of the asteroids is considerably less than that of the Moon, but that mass is broken into small pieces, each with a negligible gravitational effect, so that they would be easier to mine and refashion with techniques that would be far advanced over what they were when the first space colonies came into being.

One difficulty involved in this move is that the Sun is, to

an extent, left behind. In the asteroid belt, the apparent size and actual radiation intensity of the Sun are brought down to less than a sixth what they are in the Earth-Moon system. This might create an energy problem, but we can be reasonably sure that by the time the move to the asteroids has been made, nuclear fusion will be practical and the Sun will be dispensed with, if necessary, as an energy source— as it will have been on any settled worlds beyond Mars.

The shift from the Earth-Moon system to the asteroid belt will increase the probability that the space colonies will become starships outright, for the following reasons:

1) By the mere fact that the Sun is much farther off, much smaller, no longer the source of energy, it will bulk less importantly in the mind and consciousness of the colonists. They will have in this way already taken a step toward freeing themselves of the Solar System generally, as once their ancestors had freed themselves of the Moon and as earlier had freed themselves of Earth.

2) Having moved farther from the Sun, less energy will be required for a starship to escape from the Solar System.

Eventually, some space colony, seeing no value in circling around and around the Sun forever, will make use of some advanced propulsion system to break out of orbit and carry its structure, its contents of soil, water, air, plants, animals, and people, out into the unknown. Except for the fact that the Sun will be shrinking in apparent size and that radio contact will become steadily more difficult to maintain, until both Sun and radio contact disappear altogether, plus some small linear acceleration effect, there will be no difference whatever detectable in the colony as a result of this change in motion.

. . . But it will have become a starship; and others would undoubtedly follow.

As space colonies become starships, they will lose con-

393

tact with each other, but in the vastness of the asteroid zone (as opposed to the constricted neighborhood of the Earth-Moon system) such contact will have been minimal perhaps, in any case.

And in exchange for the loss of the Sun and of contact with other populated objects, there would come the satisfaction of curiosity, of the basic itching desire to know. Why not see what the Universe looks like? What's out there, anyway? There would also be the satisfaction of the desire for freedom. There might well be an exaltation in being at last an independent component of the Universe, free of stars and planets alike.

Nor need one, perhaps, fear the slow loss of resources through imperfect cycling (what can be perfect?). Once a starship is ready for the ultimate trip, there will be no lack of resources. They are widely spread out, but the starship can travel for several years to pick up a needed item.

The space colonists may, for instance, work their way through the comet cloud at the very rim of the Solar System, watching for one of the hundred billion comets present there in its native form as a small body of frozen volatiles. One can be found, picked up, and placed in tow, to serve as a long-time source of hydrogen to keep the fusion reactor running. Given time, and the starship will have nothing in greater profusion than time, a string of them can be picked up.

Nor need the Universe be considered empty once the comet cloud is left behind. It may take hundreds or even thousands of years to reach a star, but it is probably a mistake to assume that the Universe is made up only of stars and their immediate attendants. It happens to be only the stars that we see, because they glow brightly, but there are some indications that the Universe as a whole may be considerably more massive than the sum of the masses of the stars. At least part of the additional mass may be small, dark bodies fleeting through the volume of space between the stars.

It may be that no decade will pass without the detection of one or two of these small bodies (and occasionally a fairly large one), which may contain enough involatile material to give the starship an opportunity to flesh out its stores of aluminum, steel, and silicates.

Eventually—a long eventually, many generations, perhaps—the starship may approach a star. It might not be accident. Undoubtedly, the ship's astronomers will study all stars within so many light-years' distance, choose one that has a high probability of containing habitable worlds, and head for that.

It might be delightful to land on a world and to experience something remembered only in the mists of legend. If the planet can be burrowed into, as the Moon or Mars had once been; or if it has a compatible ecology and no dangerous life forms and can be inhabited on the outside; it may even be tempting to remain there indefinitely, abandoning the starship, which, despite all repairs, might by then be rather battered.

There would be a loss of freedom in doing so, for once again the descendants of the original planetbound humans would be planetbound, once again doomed to be perpetually pinned to the planetary surface under the light of a sun. (Yet the novelty might seem so delightful that it would be a long time before the inhabitants would come to bemoan lost freedom.)

And there would be a gain, too. Through all the centuries of starship travel, population would have had to be rigidly controlled, resources rigidly conserved and recycled. Now, for a time at least, human beings could spread over the face of a planet, glorying in explosive breeding and expansive wasting. Someday, of course, the time would come when space colonies would be formed again and these converted into starships for the renewal and continuation of the vast exploration.

We might almost imagine human societies as existing in two alternating forms: a motile, population-controlled form in starships drifting through space; and a sessile, population-expanding form on planets circling suns.

The long separation of various groups of human beings from other groups would encourage cultural and even biological variations that would produce an infinite richness of experience and culture—richness that could not conceivably be duplicated on a single world or in a single planetary system.

Competing cultures might have a chance to interact when the paths of two starships intersected.

Detecting each other from a long distance, the approach of two starships might be a time of great excitement on each. The meeting would involve a ritual of incomparable importance; there would be no flash-by with a hail and farewell.

Each, after all, would have compiled its own records, which it could now make available to the other. There would be descriptions, by each, of sectors of space never visited by the other. New theories and novel interpretations of old ones would be expounded. Literature and works of art could be exchanged, differences in custom explained.

Most of all, there would be the opportunity for a cross-flow of genes. An exchange of population (either temporary or permanent) might be the major accomplishment of any such meeting. And if the separation had been so long-enduring that the two groups had evolved into separate species, mutually infertile, there would be an intellectual cross-fertilization at any rate.

In this way, mankind would become no longer a creature of Earth or the Solar System, but of the whole Universe, drifting outward, ever outward, perhaps even out of our galaxy and into others, until such time as the Universe finally came to an enormously slow end and through one route or another could no longer support life.

But wait; we have no right to consider ourselves the only intelligent beings in the Universe. In our own galaxy there may be hundreds of millions of Earthlike planets, each with its load of life. And there may be similar numbers in each of the many billions of other galaxies. Even if only on one life-bearing planet out of millions intelligent life forms evolve and form technological civilizations, there may still be millions of such civilizations in the Universe.

Every one of these may develop to the point where (like mankind today) they stand in danger of being killed by their own success. Those who survive that danger (as we may perhaps—but not certainly—survive it) may go on, more or less inexorably, to the starship stage.

Depending on when an intelligent species got its start and how rapidly it developed, the starships may have been launched, in some cases, many millions of years ago, while others will be launched many millions of years from now.

As to those already launched, we might wonder why they have not reached us.

That may be because the vast distances of the Universe insulate us. In the entire lifetime of the galaxies so far, it may be that starships have penetrated and explored only a fantastically small percentage of the whole. Or else they may have missed us through sheer chance: passing through our sector of the Universe, they may simply never have come close enough to us to observe or be observed. Or else they may know of our existence, but choose to leave us alone as part of some basic ethic that all intelligent species have the right to develop undisturbed.

It may be, though, that on some wholly fortunate day, a human starship will encounter an alien starship. A genetic interchange would then be out of the question of course, but the intellectual cross-fertilization would be incredibly richer than in the case of a human-human interchange, assuming the human-alien transfer could be made at all.

Surely, by that point in history, it will be understood that it is the nature of the mind that makes individuals kin and that the differences in the shape, form, or manner of the material atoms out of whose intricate interrelationships that mind is built are altogether trivial.

It may be that as the human starships start moving out-ward—with what advances in human knowledge and de-velopment we can scarcely imagine—they may eventually find themselves to be part of a vast brotherhood of in-telligence, a complex of the innumerable routes by which the Universe has evolved in order to become capable of un-derstanding itself.

And as part of this brotherhood, spanning and filling the Universe, mankind will have found its true goal at last.

The Moon as Threshold

FOOTNOTES

[1] This is an enormous "if" which I by no means wish to minimize. As the crisis brought on by heedless overpopulation and criminal waste of resources gathers about the collective head of humanity, it is only too likely that we may destroy our civilization in the course of the next thirty years—and if so, this essay will merely be a glimpse into the might-have-been and nothing more.

[2] Really large meteorites would be deadly, of course, but these are excessively rare—and they would be deadly on Earth too, after all.

[3] To be sure, I am assuming here that it will be possible for human beings to be born under low-gravity conditions and to grow and develop with full physiological functioning. We won't certainly know till we try.

[4] Naturally, lunar colonists will find it far more uncomfortable to be on Earth's surface and may, in fact, never visit Earth (and feel that to be small loss).

[5] I assume here, by the way, that Mars does not, in fact, have a living ecology of its own. It may not, but we can't be sure; in fact, astronomers and biologists hope fervently that it does. And if it does, it may be ethically improper to colonize the planet.

[6] We must not overestimate the effect of isolation, however. The population of Earth itself is fearfully isolated today, since we know of no other inhabited world—not one—and yet the sense of isolation does not overwhelm us.